Studies in Computational Intelligence

Volume 1127

Series Editor

Janusz Kacprzyk, Polish Academy of Sciences, Warsaw, Poland

The series "Studies in Computational Intelligence" (SCI) publishes new developments and advances in the various areas of computational intelligence—quickly and with a high quality. The intent is to cover the theory, applications, and design methods of computational intelligence, as embedded in the fields of engineering, computer science, physics and life sciences, as well as the methodologies behind them. The series contains monographs, lecture notes and edited volumes in computational intelligence spanning the areas of neural networks, connectionist systems, genetic algorithms, evolutionary computation, artificial intelligence, cellular automata, self-organizing systems, soft computing, fuzzy systems, and hybrid intelligent systems. Of particular value to both the contributors and the readership are the short publication timeframe and the world-wide distribution, which enable both wide and rapid dissemination of research output.

Indexed by SCOPUS, DBLP, WTI Frankfurt eG, zbMATH, SCImago.

All books published in the series are submitted for consideration in Web of Science.

M. Eugenia Cornejo · László T. Kóczy ·
Jesús Medina · Eloísa Ramírez-Poussa
Editors

Computational Intelligence and Mathematics for Tackling Complex Problems 5

Springer

Editors
M. Eugenia Cornejo
Department of Mathematics
University of Cádiz
Cádiz, Spain

László T. Kóczy
Faculty of Engineering Sciences
Széchenyi István University
Győr, Hungary

Jesús Medina
Department of Mathematics
Faculty of Science
University of Cádiz
Cádiz, Spain

Eloísa Ramírez-Poussa
Department of Mathematics
University of Cádiz
Cádiz, Spain

ISSN 1860-949X ISSN 1860-9503 (electronic)
Studies in Computational Intelligence
ISBN 978-3-031-46981-7 ISBN 978-3-031-46979-4 (eBook)
https://doi.org/10.1007/978-3-031-46979-4

This Springer imprint is published by the registered company Springer Nature Switzerland AG
The registered company address is: Gewerbestrasse 11, 6330 Cham, Switzerland

Paper in this product is recyclable.

Preface

This volume is focused on connecting two interesting research areas, mathematics and computational intelligence, by means of appealing contributions devoted to give solutions to different challenges of the current technological age. Specifically, this volume is composed of the final versions of the highest quality papers presented at the conference 14th European Symposium on Computational Intelligence and Mathematics (ESCIM2022) held in Naples (Italy), from October 2nd to 5th, 2022, and that have been selected for this issue. Moreover, the technical program of this conference included three excellent keynote speeches, given by Profs. Marek Reformat (Evidence Theory Approach to Determine States of Hierarchical System), Irina Perfilieva (Fuzzy sets as manifolds, a new direction in data analysis) and Stefania Costantini (Digital Forensics: a case for Neuro-Symbolic approaches). We acknowledge the work given by the Program Committee members, reviewers, authors, local organizers and the support of the 2014–2020 ERDF Operational Programme in collaboration with the State Research Agency (AEI) in project PID2019-108991GB-I00, and with the Ecological and Digital Transition Projects 2021 of the Ministry of Science and Innovation in project TED2021-129748B-I00, the Department of Economy, Knowledge, Business and University of the Regional Government of Andalusia in project FEDER-UCA18-108612, by the National Research, Development and Innovation Office (Hungary) grant nr. K124055, and by the European Cooperation in Science and Technology (COST) Action CA17124. In the next, the papers will be briefly described.

In the first chapter "Additional Notes on Heterogeneous Concept-Forming Operators", the authors investigate a recent generalization of formal contexts considered in formal concept analysis, heterogeneous formal contexts. This kind of context allow us to diversify the data structures of objects, attributes and fuzzy relations, as a consequence they can provide a more efficient representation of data in heterogeneous environments.

The second chapter "On Measuring the Solvability of a Fuzzy Relation Equation" focuses on measuring the solvability of a fuzzy relation equation. Specifically, the authors define three mappings that assign a value of unsolvability to a multi-adjoint relation equation, making use of the conservative/pessimistic approach which transforms an unsolvable multi-adjoint relation equation into a solvable one, called solvable approximation. High values of unsolvability indicate that the approximation is not good, whereas low values mean that the multi-adjoint relation equation is close to being solvable.

The third chapter "Efficiency of Fuzzy Rough Set Decision Algorithms" studies decision algorithms given in Rough Set Theory. Decision algorithms are used in this theory to know if a set of decision rules is suitable to describe a decision table. One of the most important notions associated with decision algorithms is the notion of efficiency. In this work, the authors present a generalized notion of efficiency, considering the fuzzy framework.

The fourth chapter "Use of Fuzzy Time Series to Generate Linguistic Descriptions of Noise Pollution" introduces a proposal to represent sets of time series by means of fuzzy time series composed of linguistic labels. This representation increases the interpretability of the fuzzy time series by using linguistic labels, which can be defined by an expert in the domain. The authors use the Monte Carlo method in order to match each fuzzy set in the initial fuzzy time series with the considered linguistic labels.

The fifth chapter "A Novel Fingerprint Identification Fuzzy System Using a Center-Distance Weighted Local Binary Pattern" aims to improve the fingerprint identification system by using a center distance weighted local binary pattern. The proposed identification system begins with image enhancement using Fourier domain analysis, cutting the image around the core area. After that, the feature vectors of the local binary images are extracted and matched based on the distance of their histograms. Lastly, the retrieval of the suitable linguistic form to help clear the uncertainty of the results is made by applying a fuzzy approach.

The sixth chapter "Global Sensitivity Analysis and Low Magnitude Pruning for Convolutional Neural Networks Reduction in Image Net Based on Transfer Learning State of the Art Models" investigates the Global Sensitivity Analysis method for creating generalized pruned networks for transfer learning. The goal is to discern whether the Global Sensitivity Analysis method can be competitive with well-established methods which also reduce popular convolutional neural networks.

For that reason, the authors carry out a comparative study between the Global Sensitivity Analysis method and the Low Magnitude Pruning method, by using experimental data.

The seventh chapter "Generalized Linearity of Aggregation and Related Functions" studies aggregation functions and some of their generalizations which are linear splines. The approach addressed in this work provides an alternative definition of the Choquet integral, as well as a parametric class of singular copulas with pre-described support.

The authors of the eighth chapter "Closure Operators on AMV-Valued Fuzzy Sets" tackle the problem of transforming L-fuzzy structures with values in a complete MV-algebra L, into $F(L)$-fuzzy sets $X \to F(L)$, where F is a functor from the category of MV-algebras to the category of the so-called almost MV-algebras. The author shows how this transformation can be used to define closure operators on the powerset objects of these new *L*-fuzzy structures.

The ninth chapter "An Application of a Fuzzy Multi-criteria Decision Making Process for Explainable Machine Learning in the Actuarial Context" presents a case study of applying Fuzzy Multi-Criteria Decision-Making (MCDM) models for solving actuarial problems using machine learning methods. In particular, these models are applied to the problem of pricing in health insurance.

The tenth chapter "A Software Tool for Solving the Traveling Salesman Problem and Related Non-fuzzy and Fuzzy Optimization Problems" develops a software tool for solving the Traveling Salesman Problem and closely related non-fuzzy and fuzzy optimization problems. The Discrete Bacterial Memetic Evolutionary Algorithm is implemented in the software since it is an efficient method for solving the for handling the examined type of optimization problems and their extensions.

In the eleventh chapter "Characterizing Clarify and Reduce Method by Means of Left-Sided Formal Concept Analysis", the authors explore reduction techniques of contexts in formal concept analysis, within the multi-adjoint paradigm. In this paper, the relation between the reduction mechanism considered in the multi-adjoint framework and the Clarify-and-Reduce-method has been studied.

The twelfth chapter "ITAMACDSS: A Fuzzy Clinical Diagnostic Support System for Coeliac Disease" proposes a fuzzy clinical diagnostic support system for coeliac disease in children. The author presents and discusses the obtained results after applying the diagnostic support system on an unprecedented data collection on more than 20K children, both in Malta and Sicily.

The thirteenth chapter "Preference in Multi-adjoint Logic Programming Based on Ordered Adjoint Pairs" analyzes how the choice of different ordered adjoint pairs can affect the computation of the minimal model of a fuzzy multi-adjoint logic program, with the primary goal of studying preferences in decision-making frameworks.

The fourteenth chapter "Lattice Valued Aggregation Operators Applicable in Digital Forensics" presents a mathematical tool, based on the use of bounded lattices and aggregation operators, capable of ranking data in different problems of practical areas. The research also shows how to obtain a suitable measure of interaction that reflects the opinions of experts in digital forensics.

The fifteenth chapter "Power Consumption-Based Identification of Used Encryption Solution in IoT Environments" proposes an approach for identifying the used protocol and encryption solutions in network communications, based on power consumption and energy efficiency. Different Nvidia Jetson devices, in different power modes and protocols, are analyzed. The reliability of the proposed approach is determined by using classification algorithms.

The sixteenth chapter "Maritime Cyber-Attacks Detection Based on a Convolutional Neural Network" addresses the cybersecurity in maritime communications. The author designs a convolutional neural network for the identification of ship types and the detection of anomalous GNSS data received by ships, which are being targeted by cyber-attackers.

The authors of the seventeenth chapter "Epistemic Logic and Theory of Mind for Modelling Group Dynamics of Criminal Organizations" show how the Logic of "Inferable" L-DINF is able to model (aspects of) the group dynamics of cooperative agents. They explain how this logic allows a designer to model real-world situations encompassing joint intentions and plans with roles, preferences and costs concerning action execution. This investigation may be of special interest for the investigation of the activities of criminal organizations.

The eighteen chapter "Effective Contract-Net-Based Intrusion Detection Using Intelligent Agent-Based Systems" addresses the problem of intrusion detection. This work proposes an intelligent agent security approach for adopting Intrusion Detection Systems in a distributed computational network-based environment, which provides advantages such as scalability or no single point of failure.

In the last chapter "Analysing the Influence of the Pandemic on Crime Patterns in North Macedonia", machine learning is used to analyze the impact of the pandemic of COVID-19 on the crime rates in the case of the study of North Macedonia.

The authors apply classification, time series analysis and forecasting to extract the past and current trends in crime rates.

Cádiz, Spain M. Eugenia Cornejo
Győr, Hungary László T. Kóczy
Cádiz, Spain Jesús Medina
Cádiz, Spain Eloísa Ramírez-Poussa

Contents

Additional Notes on Heterogeneous Concept-Forming Operators

Peter Eliaš, Ľubomír Antoni, Ondrej Krídlo, and Stanislav Krajči

1 Introduction

Definition of L-fuzzy concept [1] as a generalization of a formal concept [2], induced similarities [3], fuzzy Galois connections [4, 5], scaling of formal context [6], or the algorithm based on the pruning of the fuzzy concepts lattice [7] were the first results connected to the so-called Fuzzy Formal Concept Analysis. The one-sided fuzzifications of formal contexts were proposed independently in [7–9]. As the natural continuation of these approaches, Medina, Ojeda-Aciego and Vojtáš [10–12] explored a multi-adjoint approach, which was fruitfully applied in Formal Concept Analysis [13–19], as well.

Our generalizations in Fuzzy Formal Concept Analysis include the diversification of the types of objects, attributes, or fuzzy relations [20–22]. Moreover, Pócs proposed a construction of fuzzy concept lattices via Galois connections [23–25], which was later extended and explored in [26–29]. The cooperative conversional recommendation [30], supervised classification performance analysis [31], disjunctive attribute dependencies [32], fuzzy relational equations [33], or OWA operators [34]

P. Eliaš
Mathematical Institute, Slovak Academy of Sciences, Grešákova 6, 040 01 Košice, Slovak Republic
e-mail: elias@saske.sk

L. Antoni (✉) · O. Krídlo · S. Krajči
Institute of Computer Science, Faculty of Science, Pavol Jozef Šafárik University in Košice, 041 80 Košice, Slovakia
e-mail: lubomir.antoni@upjs.sk

O. Krídlo
e-mail: ondrej.kridlo@upjs.sk

S. Krajči
e-mail: stanislav.krajci@upjs.sk

M. E. Cornejo et al. (eds.), *Computational Intelligence and Mathematics for Tackling Complex Problems 5*, Studies in Computational Intelligence 1127,
https://doi.org/10.1007/978-3-031-46979-4_1

are other interesting issues which are explored in the connection with Fuzzy Formal Concept Analysis.

Heterogeneous formal context [22, 35] in Formal Concept Analysis represents our recent generalization of formal context [2] which allows us to diversify the data structures of each object, each attribute and each value of fuzzy relation. In the paper, we aim to extend our previous results regarding the heterogeneous concept-forming operators.

2 Heterogeneous Concept-Forming Operators

In our recent papers [22, 35], we proposed the notions of heterogeneous formal context and its decomposition, heterogeneous concept-forming operators, heterogeneous formal concepts, and heterogeneous concept lattices. We remind only the most important notions here.

Let A be a set of objects and B be a set of attributes. Consider the systems of complete lattices $\mathcal{C} = ((C_a, \leq_{C_a}) : a \in A)$ and $\mathcal{D} = ((D_b, \leq_{D_b}) : b \in B)$. Let F be a set of all $\mathcal{C}$-fuzzy subsets of A. Let G be a set of all $\mathcal{D}$-fuzzy subsets of B. Finally, consider $\mathcal{P} = ((P_{b,a}, \leq_{P_{b,a}}) : b \in B, a \in A)$ as a system of partially ordered sets and R as a $\mathcal{P}$-fuzzy relation on $B \times A$ (for more details see [22, 35]).

We consider system of operations $(\bullet_{b,a} : b \in B, a \in A)$ such that for all $b \in B, a \in A$ we have $\bullet_{b,a} : D_b \times C_a \to P_{b,a}$ and it holds that $\bullet_{b,a}$ is isotone and left-continuous in both arguments, i.e.,

(1a) If $c_1 \leq c_2$, then $d \bullet_{b,a} c_1 \leq d \bullet_{b,a} c_2$ for $d \in D_b$ and for all $c_1, c_2 \in C_a$,
(1b) If $d_1 \leq d_2$, then $d_1 \bullet_{b,a} c \leq d_2 \bullet_{b,a} c$ for all $d_1, d_2 \in D_b$ and $c \in C_a$,
(2a) If $d \bullet_{b,a} c \leq p$ for some $d \in D_b$, $p \in P_{b,a}$ and for all $c \in X \subseteq C_a$, then $d \bullet_{b,a} \sup X \leq p$, and
(2b) If $d \bullet_{b,a} c \leq p$ for some $c \in C_a$, $p \in P_{b,a}$ and for all $d \in Y \subseteq D_b$, then $\sup Y \bullet_{b,a} c \leq p$.

Now, we define the mappings $\nearrow : G \to F$ and $\swarrow : F \to G$. For $g \in G$, $\nearrow(g) \in F$ is given by

$$(\nearrow(g))(a) = \sup\{c \in C_a : (\forall b \in B) g(b) \bullet_{b,a} c \leq R(b,a)\} \text{ for all } a \in A.$$

For $f \in F$, $\swarrow(f) \in G$ is given by

$$(\swarrow(f))(b) = \sup\{d \in D_b : (\forall a \in A) d \bullet_{b,a} f(a) \leq R(b,a)\} \text{ for all } b \in B.$$

The full version of definitions and properties of heterogeneous formal context and its concept lattice can be found in [22, 35].

3 Extended Results

In this section, we will extend our results regarding the heterogeneous concept-forming operators.

Note that $F = \prod_{a \in A} C_a$ (with the projections $\pi_a : F \to C_a$) is categorical product of lattices C_a. This holds for category of complete lattices **CLat** (the morphisms are represented by complete lattice morphisms), the category of lattices **Lat** (morphisms are represented by lattice morphisms), the category of posets **Pos** (morphisms are represented by mappings which preserve ordering), the category **SupLat** (morphisms are represented by mappings which preserve supremum), or the category **InfLat** (morphisms are represented by mappings which preserve infimum).

It means that F is a complete lattice since C_a is a complete lattice for all $a \in A$. Moreover, each system of morphisms from a complete lattice L to C_a can be uniquely represented as one morphism from L to F. Analogously can be considered a set of all $\mathcal{D}$-fuzzy subsets of B which is denoted by G.

Regarding the definition of supremum in the complete lattices G and F, the mappings $\nearrow : G \to F$, $\swarrow : F \to G$ can be equivalently expressed in the following way:

$$\nearrow(g) = \bigvee \{f \in F : (\forall b \in B)(\forall a \in A)\, g(b) \bullet_{b,a} f(a) \leq R(b, a)\},$$
$$\swarrow(f) = \bigvee \{g \in G : (\forall b \in B)(\forall a \in A)\, g(b) \bullet_{b,a} f(a) \leq R(b, a)\}.$$

Now, denote $P = \prod_{(b,a) \in B \times A} P_{b,a}$. For all $p, q \in P$, we can define the ordering $\leq$ such that

$$p \leq q$$

iff

$$(\forall b \in B)(\forall a \in A)\, p(b, a) \leq q(b, a).$$

It holds that $R \in P$. Moreover, we define the mapping $\bullet : G \times F \to P$ by

$$(g \bullet f)(b, a) = g(b) \bullet_{b,a} f(a) \quad \text{for all } b \in B, a \in A.$$

Thus, the definitions of concept-forming operators $\nearrow$ and $\swarrow$ are the following:

$$\nearrow(g) = \bigvee \{f \in F : g \bullet f \leq R\},$$
$$\swarrow(f) = \bigvee \{g \in G : g \bullet f \leq R\}.$$

Without loss of generality, we can suppose that P is a complete lattice.

The properties (1a), (1b), (2a), (2b) of system of operations $\bullet_{b,a}$ can be equivalently formulated by the mapping $\bullet$:

(3) if $g_1, g_2 \in G$, $f_1, f_2 \in F$, $g_1 \leq g_2$ and $f_1 \leq f_2$, then $g_1 \bullet f_1 \leq g_2 \bullet f_2$,
(4) if for $G' \subseteq G$, $F' \subseteq F$ and $p \in P$ holds that $(\forall g \in G')(\forall f \in F')\, g \bullet f \leq p$, then $\left(\bigvee G'\right) \bullet \left(\bigvee F'\right) \leq p$.

If P is a complete lattice, then the pair of properties (3) and (4) are equivalent to:

(5) $\bigvee\{g \bullet f : g \in G',\ f \in F'\} = \left(\bigvee G'\right) \bullet \left(\bigvee F'\right)$ for all $G' \subseteq G$ and $F' \subseteq F$.

Moreover, property (3) is equivalent to the pair of properties:

(3a) if $g \in G$, $f_1, f_2 \in F$ and $f_1 \leq f_2$, then $g \bullet f_1 \leq g \bullet f_2$,
(3b) if $g_1, g_2 \in G$, $f \in F$ and $g_1 \leq g_2$, then $g_1 \bullet f \leq g_2 \bullet f$.

Property (4) is equivalent to the pair of properties:

(4a) if for $g \in G$, $F' \subseteq F$, $p \in P$ holds that $(\forall f \in F')\, g \bullet f \leq p$, then $g \bullet \left(\bigvee F'\right) \leq p$,
(4b) if for $G' \subseteq G$, $f \in F$, $p \in P$ holds that $(\forall g \in G')\, g \bullet f \leq p$, then $\left(\bigvee G'\right) \bullet f \leq p$.

Property (5) is equivalent to the pair of properties:

(5a) for all $g \in G$ and $F' \subseteq F$ holds that $\bigvee\{g \bullet f : f \in F'\} = g \bullet \left(\bigvee F'\right)$,
(5b) for all $G' \subseteq G$ and $f \in F$ holds that $\bigvee\{g \bullet f : g \in G'\} = \left(\bigvee G'\right) \bullet f$.

In the following, we will assume that P is a complete lattice. From the properties (5a), (5b) we have that $\bullet\colon G \times F \to P$ is the so-called bimorphism in SupLat, which means that the mappings $\phi_g : F \to P$ and $\phi^f : G \to P$ defined by $\phi_g(f) = \phi^f(g) = g \bullet f$ preserve the arbitrary suprema for all $g \in G$ and $f \in F$. It means that they are the morphisms in SupLat.

Note that $G \times F$ is a complete lattice, this the object of SupLat. Each morphism $G \times F \to P$ in SupLat is a bimorphism in SupLat, as well. It does not hold in an opposite way. For example, if $G = F = P = \{0, 1\}$, then $\phi(g, f) = \min\{g, f\}$ is a bimorphism, but not a morphism.

The following theorems show the relationship between a bimorphism $\phi\colon G \times F \to P$ and the pair of mappings $\nearrow$ and $\swarrow$.

Theorem 1 *Let F, G, P be the complete lattices, let $\phi\colon G \times F \to P$ be a bimorphism in* SupLat *and $p \in P$, the the pair of mappings*

$$\nearrow(g) = \bigvee\{f \in F : \phi(g, f) \leq p\}, \quad \swarrow(f) = \bigvee\{g \in G : \phi(g, f) \leq p\} \tag{6}$$

forms a Galois connection from G to F.

Theorem 2 *Let F, G, P be the complete lattices. Let $\phi\colon G \times F \to P$ be a bimorphism such that if $\langle\nearrow, \swarrow\rangle$ is a Galois connection from G to F, then the mappings $\nearrow, \swarrow$ are defined by*

$$\nearrow(g) = \bigvee\{f \in F : \phi(g, f) \leq p\}, \quad \swarrow(f) = \bigvee\{g \in G : \phi(g, f) \leq p\}$$

for $p \in P$.

In particular, P *is a complete lattice of all Galois connections from* G *to* F *ordered by relation* $\langle \nearrow_1, \swarrow_1 \rangle \leq \langle \nearrow_2, \swarrow_2 \rangle$ *iff* $(\forall g \in G) \nearrow_1(g) \leq \nearrow_2(g)$. *A bimorphism* ϕ *is defined by* $\phi(g, f) = \bigwedge\{\langle \nearrow, \swarrow \rangle \in P : f \leq \nearrow(g)\}$ *for* $g \in G$ *and* $f \in F$, *and* p *is a Galois connection* $\langle \nearrow, \swarrow \rangle$.

The proofs of these theorems will be added in the extended version of paper (due to limited pages for this paper). The proof of Theorem 2 is based on the notion of join-reversing mappings [36]. Moreover, we will provide the examples of introduced notions in our extended version of paper.

4 Conclusions and Future Work

In this paper, we presented some additional results regarding heterogeneous formal context and its heterogeneous concept-forming operators. We described the relationship between a particular bimorphism and the pair of concept-forming operators. The proofs of the theorems and the examples of introduced notions will be supplemented in the extended version of paper.

Acknowledgements This work was supported by the Slovak Research and Development Agency under contract No. APVV-21-0468 (Ľubomír Antoni, Stanislav Krajči, Ondrej Krídlo) and by the Scientific Grant Agency of the Ministry of Education, Science, Research and Sport of the Slovak Republic and Slovak Academy of Sciences under contract VEGA 2/0097/20 (Peter Eliaš).

References

1. Burusco, A., Fuentes-González, R.: The study of L-fuzzy concept lattice. Mathw. Soft Comput. **3**, 209–218 (1994)
2. Ganter, G., Wille, R.: Formal Concept Analysis, Mathematical Foundation. Springer, Berlin (1999)
3. Bělohlávek, R.: Fuzzy concepts and conceptual structures: induced similarities. In: JCIS '98 proceedings, International Conference on Computer Science and Informatics, pp. 179–182. Association for Intelligent Machinery (1998)
4. Bělohlávek, R.: Fuzzy Galois connections. Math. Log. Q. **45**(4), 497–504 (1999)
5. Bělohlávek, R.: Concept lattices and order in fuzzy logic. Ann. Pure Appl. Logic **128**, 277–298 (2004)
6. Pollandt, S.: Datenanalyse mit Fuzzy-Begriffen. In: G. Stumme, R. Wille (eds.), Begriffliche Wissensverarbeitung. Methoden und Anwendungen, pp. 72–98. Springer, Heidelberg (2000)
7. Ben Yahia, S., Jaoua, A.: Discovering knowledge from fuzzy concept lattice. In: Kandel, A., Last, M., Bunke, H. (eds.) Data Mining and Computational Intelligence, pp. 169–190. Physica-Verlag (2001)
8. Bělohlávek, R., Sklenář, V., Zacpal, J.: Crisply generated fuzzy concepts. Lect. Notes Comput. Sci. **3403**, 268–283 (2005)
9. Krajči, S.: Cluster based efficient generation of fuzzy concepts. Neural Netw. World **13**, 521–530 (2003)

10. Medina, J., Ojeda-Aciego, M., Valverde, A., Vojtáš, P.: Towards biresiduated multi-adjoint logic programming. Lect. Notes Artif. Intell. **3040**, 608–617 (2004)
11. Medina, J., Ojeda-Aciego, M., Vojtáš, P.: Multi-adjoint logic programming with continuous semantics. Lect. Notes Artif. Intell. **2173**, 351–364 (2001)
12. Medina, J., Ojeda-Aciego, M., Vojtáš, P.: Similarity-based unification: a multi-adjoint approach. Fuzzy Sets Syst. **146**, 43–62 (2004)
13. Cornejo, M.E., Medina, J., Ramírez, E.: A comparative study of adjoint triples. Fuzzy Sets Syst. **211**, 1–14 (2013)
14. Cornejo, M.E., Medina, J., Ramírez, E.: Characterizing reducts in multi-adjoint concept lattices. Inf. Sci. **422**, 364–376 (2018)
15. Madrid, N., Ojeda-Aciego, M.: Multi-adjoint lattices from adjoint triples with involutive negation. Fuzzy Sets Syst. **405**, 88–105 (2021)
16. Medina, J., Ojeda-Aciego, M.: Multi-adjoint t-concept lattices. Inf. Sci. **180**, 712–725 (2010)
17. Medina, J., Ojeda-Aciego, M.: On multi-adjoint concept lattices based on heterogeneous conjunctors. Fuzzy Sets Syst. **208**, 95–110 (2012)
18. Medina, J., Ojeda-Aciego, M., Ruiz-Calviño, J.: Formal concept analysis via multi-adjoint concept lattices. Fuzzy Sets Syst. **160**, 130–144 (2009)
19. Medina, J., Ojeda-Aciego, M., Pócs, J., Ramírez-Poussa, E.: On the Dedekind-MacNeille completion and formal concept analysis based on multilattices. Fuzzy Sets Syst. **303**, 1–20 (2016)
20. Krajči, S.: A generalized concept lattice. Logic J. IGPL **13**, 543–550 (2005)
21. Krídlo, O., Krajči, S., Antoni, L'.: Formal concept analysis of higher order. Int. J. Gen. Syst. **45**(2), 116–134 (2016)
22. Antoni, L.', Krajči, S., Krídlo, O., Macek, B., Pisková, L.: On heterogeneous formal contexts. Fuzzy Sets Syst. **234**, 22–33 (2014)
23. Pócs, J.: Note on generating fuzzy concept lattices via Galois connections. Inf. Sci. **185**, 128–136 (2012)
24. Pócs, J.: On possible generalization of fuzzy concept lattices using dually isomorphic retracts. Inf. Sci. **210**, 89–98 (2012)
25. Pócs, J., Pócsová, J.: Basic theorem as representation of heterogeneous concept lattices. Front. Comput. Sci. **9**(4), 636–642 (2015)
26. Butka, P., Pócs, J.: Generalization of one-sided concept lattices. Comput. Inform. **32**(2), 355–370 (2013)
27. Butka, P., Pócs, J., Pócsová, J.: Representation of fuzzy concept lattices in the framework of classical FCA. J. Appl. Math. ID236725 (2013)
28. Butka, P., Pócs, J., Pócsová, J.: Distributed computation of generalized one-sided concept lattices on sparse data tables. Comput. Inform. **34**(1), 77–98 (2015)
29. Halaš, R., Pócs, J.: Generalized one-sided concept lattices with attribute preferences. Inf. Sci. **303**, 50–60 (2015)
30. Cordero, P., Enciso, M., Mora, Á., Ojeda-Aciego, M., Rossi, C.: A formal concept analysis approach to cooperative conversational recommendation. Int. J. Comput. Intell. Syst. **13**(1), 1243–1252 (2020)
31. Valverde-Albacete, F.J., Peláez-Moreno, C.: A framework for supervised classification performance analysis with information-theoretic methods. IEEE Trans. Knowl. Data. Eng. **32**(11), 2075–2087 (2020)
32. Dubois, D., Medina, J., Prade, H., Ramírez-Poussa, E.: Disjunctive attribute dependencies in formal concept analysis under the epistemic view of formal contexts. Inf. Sci. **561**, 31–51 (2021)
33. Medina, J.: Minimal solutions of generalized fuzzy relational equations: clarifications and corrections towards a more flexible setting. Int. J. Approx. Reason. **84**, 33–38 (2017)
34. Medina, J., Yager, R.R.: OWA operators with functional weights. Fuzzy Sets Syst. **414**, 38–56 (2021)
35. Antoni, L'., Eliaš, P., Krajči, S., Krídlo, O.: Heterogeneous formal context and its decomposition by heterogeneous fuzzy subsets. Fuzzy Sets Syst. Article in Press (2022). https://doi.org/10.1016/j.fss.2022.05.015

36. Höhle U.: Modules in the category Sup. In: Saminger-Platz, S., Mesiar, R. (eds.) On Logical, Algebraic, and Probabilistic Aspects of Fuzzy Set Theory, Vol. 336 of Studies in Fuzziness and Soft Computing. Springer, Berlin (2016)

On Measuring the Solvability of a Fuzzy Relation Equation

David Lobo, Víctor López-Marchante, and Jesús Medina

1 Introduction

Fuzzy relation equations (FRE) have gained relevance in the field of approximate reasoning in the last few years. They have applications in several areas as optimization [1], image processing [5] and abductive reasoning [9, 10]. As it is clear from different works on the resolution of FRE [6, 14], one of the biggest problems that usually emerges is the lack of solutions, i.e., their unsolvability.

The reason behind the unsolvability of a FRE might be imprecision associated with the data extraction or the existence of inconsistencies or incoherences in the data. As a result, trying to measure the solvability of an unsolvable FRE is a crucial matter.

In this paper, we will consider multi-adjoint relation equations (MARE) [7], because of its flexibility, although the study carried out can be applied to other FRE frameworks [2, 6, 8, 13]. We will define three mappings that assign a value of unsolvability to a MARE, depending on how far it is from being solvable. To this aim, we will make use of the *conservative/pessimistic* approach introduced in [3], which

Supported by the 2014–2020 ERDF Operational Programme in collaboration with the State Research Agency (AEI) in projects PID2019-108991GB-I00 and PID2022-137620NB-I00, and with the Department of Economy, Knowledge, Business and University of the Regional Government of Andalusia in project FEDER-UCA18-108612, and by the European Cooperation in Science & Technology (COST) Action CA17124.

D. Lobo (✉) · V. López-Marchante · J. Medina
Department of Mathematics, University of Cádiz, Cádiz, Spain
e-mail: david.lobo@uca.es

V. López-Marchante
e-mail: victor.lopez@uca.es

J. Medina
e-mail: jesus.medina@uca.es

M. E. Cornejo et al. (eds.), *Computational Intelligence and Mathematics for Tackling Complex Problems 5*, Studies in Computational Intelligence 1127,
https://doi.org/10.1007/978-3-031-46979-4_2

transforms an unsolvable MARE into a solvable one, called solvable approximation, by modifying its right-hand side. High values of unsolvability will mean that the approximation is not good and low values will mean that the MARE is close to be solvable.

The rest of the paper is organized as follows. Section 2 shows some notions and results concerning MARE, as well as a method for the approximation of unsolvable MARE. Then, three forms of measuring the solvability of a MARE are introduced in Sect. 3. Finally, in Sect. 4, we discuss about different types of unsolvability according to two toy examples. We finish with some conclusions and prospects for future work.

2 Approximation of Unsolvable Multi-adjoint Relation Equations

In this section, some notions related to MARE and the approximation method presented in [3] will be recalled. First of all, we recall the underlying algebraic structure of a MARE, which relies on adjoint triples.

Adjoint triples can be used to define a Modus Ponens in a fuzzy framework, where the mapping representing a conjunction, denoted as &, is not necessarily commutative nor associative. As a result, this gives rise to two residuated implications that might be different.

Definition 1 Let [4] $(P_1, \preceq_1)$, $(P_2, \preceq_2)$, $(P_3, \preceq_3)$ be posets and $\&\colon P_1 \times P_2 \to P_3$, $\swarrow\colon P_3 \times P_2 \to P_1$, $\nwarrow\colon P_3 \times P_1 \to P_2$ mappings, then $(\&, \swarrow, \nwarrow)$ is an *adjoint triple* with respect to P_1, P_2, P_3 if

$$x \preceq_1 z \swarrow y \quad \text{iff} \quad x \& y \preceq_3 z \quad \text{iff} \quad y \preceq_2 z \nwarrow x$$

for each $x \in P_1$, $y \in P_2$, $z \in P_3$.

The main foundation of the multi-adjoint paradigm is the allowance of different adjoint triples in a single framework. In this case, we are interested in multi-adjoint property-oriented frames [12].

Definition 2 Let [12] $(L_1, \preceq_1)$, $(L_2, \preceq_2)$ be two lattices, $(P, \preceq_3)$ a poset and $\{(\&_i, \swarrow^i, \nwarrow_i) \mid i \in \{1, \ldots, n\}\}$ a set of adjoint triples with respect to P, L_2, L_1. The tuple

$$(L_1, L_2, P, \preceq_1, \preceq_2, \preceq_3, \&_1, \swarrow^1, \nwarrow_1, \ldots, \&_n, \swarrow^n, \nwarrow_n)$$

is called *multi-adjoint property-oriented frame*.

Essentially, a FRE is an equality of the form $R \odot S = T$, where R, S and T are fuzzy relations, $\odot$ is a sup-composition operator and either R or S is unknown. MARE follow the same idea, but considering a composition operator with a multi-adjoint nature.

Definition 3 Let [7] U, V, W be sets, $(L_1, L_2, P, \&_1, \dots, \&_n)$ a multi-adjoint property-oriented frame, $R \in P^{U\times V}$, $T \in L_1^{U\times W}$ two fuzzy relations and $\sigma\colon V \to \{1, \dots, n\}$. A *MARE with sup-$\&_\sigma$ composition* is an equality of the form

$$R \odot_\sigma X = T \tag{1}$$

where X is an unknown fuzzy relation, where the operator $\odot_\sigma\colon P^{U\times V} \times L_2^{V\times W} \to L_1^{U\times W}$ is defined as

$$R \odot_\sigma S(u, w) = \bigvee\{R(u, v)\&_{\sigma(v)}S(v, w) \mid v \in V\} \tag{2}$$

We say that a MARE is *solvable* if it admits at least one solution. Otherwise, we say that it is *unsolvable*.

Given a MARE, if we denote $L_1^U = \{f \mid f\colon U \to L_1\}$ and $L_2^V = \{g \mid g\colon V \to L_2\}$, the mappings ${}^{\uparrow_\pi}\colon L_2^V \to L_1^U$ and ${}^{\downarrow^N}\colon L_1^U \to L_2^V$, defined as

$$g^{\uparrow_\pi}(a) = \bigvee_1 \left\{R(a, b) \;\&_{\sigma(a,b)}\; g(b) \mid b \in B\right\}$$

$$f^{\downarrow^N}(b) = \bigwedge_2 \left\{f(a) \nwarrow_{\sigma(a,b)} R(a, b) \mid a \in A\right\}$$

form an isotone Galois connection [12]. As shown in the mentioned paper, we can then define a multi-adjoint property-oriented concept lattice, which will be denoted as $\mathcal{M}_{\pi N}(U, V, R, \sigma)$ or simply $\mathcal{M}_{\pi N}$ when there is no room for confusion. Among others, the concept lattice $\mathcal{M}_{\pi N}$ can be used to compute the solution set of a MARE, as well as to characterize its solvability, as the following proposition shows.

Proposition 1 *Let [7] $R \odot_\sigma X = T$ be a MARE. Then $R \odot_\sigma X = T$ is solvable if and only if $T_w = T_w^{\downarrow^N\uparrow_\pi}$ for all $w \in W$, where $T_w(u) = T(u, w)$, for all $u \in U$, $w \in W$.*

According to the characterization presented in Proposition 1, a method for transforming an unsolvable MARE into a solvable MARE was introduced in [3]. Basically, the idea of the method consists of replacing the term T_w by $T_w^{\downarrow^N\uparrow_\pi}$, which clearly verifies that

$$(T_w^{\downarrow^N\uparrow_\pi})^{\downarrow^N\uparrow_\pi} = T_w^{\downarrow^N\uparrow_\pi}$$

Hence, the resulting MARE is solvable. Since $T_w^{\downarrow^N\uparrow_\pi} \preceq T_w$, the modified MARE is known as *conservative/pessimistic approximation*.

3 Measuring the Solvability of a MARE

The procedure presented in [3] provides a solvable approximation of any MARE. Nevertheless, we have no information about how different is the original MARE from the approximation, or equivalently, how far is the term T_w from the term $T_w^{\downarrow^N \uparrow_\pi}$. At first, some MARE might need a big change in order to turn solvable, whilst small changes in the right-hand side of other MARE will guarantee their solvability. The key point behind this is the amount of inconsistent information contained in a MARE.

In this section, we study the solvability of a MARE following the philosophy of the measures presented in [11] for the coherence of extended residuated logic programs. Specifically, we present three different forms of measuring solvability which assign a value in the unit interval to a given MARE, which represents its unsolvability. From now on, a MARE

$$R \odot_\sigma X = T \qquad (\mathcal{E})$$

will be fixed. A first possibility to measure the solvability of ($\mathcal{E}$) is counting the number of sub-equations (rows) whose right-hand side has to be modified in order to achieve solvability. For instance, we can identify the modified rows by means of the mapping $I\colon U \times W \to \{0, 1\}$ defined, for all $u \in U$, $w \in W$, as:

$$I(u, w) = \begin{cases} 1 & \text{if } T_w(u) = T_w^{\downarrow^N \uparrow_\pi}(u) \\ 0 & \text{if } T_w(u) \neq T_w^{\downarrow^N \uparrow_\pi}(u) \end{cases}$$

Thus, the first form of measuring the solvability of ($\mathcal{E}$) is defined as

$$\mathcal{M}_1(\mathcal{E}) = \frac{\sum_{u \in U} \sum_{w \in W} I(u, w)}{|U|\,|W|}$$

Notice that, the mapping $\mathcal{M}_1$ just counts the number of sub-equations that change from T_w to $T_w^{\downarrow^N \uparrow_\pi}$. However, we have no knowledge about if the change in the right-hand side of the sub-equations has been noticeable or only a small adjustment has been carried out. In other words, $\mathcal{M}_1$ does not take into account the magnitude of the inconsistencies that produce the unsolvability of ($\mathcal{E}$).

In order to capture this issue, we can define a mapping $m\colon L_1 \times L_1 \to \mathbb{R}$ to measure the unsolvability of a sub-equation, where the first argument represents the right-hand side of the sub-equation in ($\mathcal{E}$) and the second argument is the right-hand side of its conservative approximation.

Clearly, the mapping m should verify the following conditions:

(a) $m(t, t) = 0$, that is, the inconsistency of any consistent sub-equation is 0.
(b) m is increasing on the first argument and decreasing on the second argument. Since $T_w^{\downarrow^N \uparrow_\pi} \preceq T_w$, this property assures that bigger changes in the right-hand side will lead to greater unsolvability values.

Given a mapping m satisfying (a) and (b) and focusing on the maximum size of unsolvability in the sub-equations of a MARE, we can define a second form of measuring solvability as follows:

$$\mathcal{M}_2(\mathcal{E}) = \sup\{m(T_w(u), T_w^{\downarrow^N\uparrow_\pi}(u)) \mid u \in U, w \in W\}$$

Notice that, the measure $\mathcal{M}_2$ is well defined since any subset of $\mathbb{R}$ has a supremum.

Finally, a third measure is introduced, under the assumptions of L_1 being a bounded lattice and m a bounded mapping. These conditions ensure the existence of a maximum unsolvability. In particular, the maximum possible unsolvability of a sub-equation is $m(\top, \bot)$. The purpose of the mapping $\mathcal{M}_3$ is estimating the average size of unsolvability, which can also be interpreted as the proportion of unsolvability in the sub-equations of a MARE.

$$\mathcal{M}_3(\mathcal{E}) = \frac{\sum_{u\in U}\sum_{w\in W} m(T_w(u), T_w^{\downarrow^N\uparrow_\pi}(u))}{|U||W|\, m(\top, \bot)}$$

4 Different Types of Unsolvability

The unsolvability of a MARE is not always due to the same kind of inconsistency. In this section, by means of two toy examples, we show how the measures $\mathcal{M}_1$, $\mathcal{M}_2$ and $\mathcal{M}_3$ can be applied to detect two different types of unsolvability. Both examples are developed in the multi-adjoint frame

$$([0, 1]_{10}, [0, 1]_{10}, [0, 1]_{10}, \overline{\&}_p)$$

where $[0, 1]_{10}$ denotes the regular partition of the unit interval in 10 pieces and $\overline{\&}_p$ is the discretization of the product t-norm.

Example 1 *Unsolvability due to minor errors in data.* Consider the unsolvable MARE given by $R \odot_\sigma X = T$ where

$$R = \begin{pmatrix} 0.4 & 0.2 \\ 0.3 & 0.6 \\ 0.7 & 0.5 \end{pmatrix}, \qquad T = \begin{pmatrix} 0.2 \\ 0.3 \\ 0.2 \end{pmatrix}$$

Carrying out the corresponding computations, we obtain $T^{\downarrow^N\uparrow_\pi} = (0.1, 0.3, 0.2)$. Hence, it is sufficient to decrease the independent term of the first sub-equation by 0.1 to achieve solvability. The solvability of the MARE is related to the following three values:

$$\mathcal{M}_1(\mathcal{E}) = \frac{1}{3} \qquad \mathcal{M}_2(\mathcal{E}) = \frac{1}{10} \qquad \mathcal{M}_3(\mathcal{E}) = \frac{1}{30}$$

As a result, $\mathcal{M}_1(\mathcal{E})$ implies that the unsolvability of the MARE is present in a third of its sub-equations, and from $\mathcal{M}_2(\mathcal{E})$ and $\mathcal{M}_3(\mathcal{E})$ we deduce that the unsolvability can be easily avoided with the conservative approximation, as it is probably due to some noise in the data.

Example 2 *Unsolvability due to the existence of unsolvable equations.* Consider the unsolvable MARE given by $R \odot_\sigma X = T$ where

$$R = \begin{pmatrix} 0.3 & 0.1 \\ 0.3 & 0.5 \\ 0.3 & 0.4 \end{pmatrix}, \qquad T = \begin{pmatrix} 0.3 \\ 0.3 \\ 1 \end{pmatrix}$$

The computation of the conservative approximation of this MARE results in $T_{w}^{\downarrow^N \uparrow_\pi} = (0.3, 0.3, 0.3)$, and its solvability is associated with the following three values:

$$\mathcal{M}_1(\mathcal{E}) = \frac{1}{3} \qquad \mathcal{M}_2(\mathcal{E}) = \frac{7}{10} \qquad \mathcal{M}_3(\mathcal{E}) = \frac{7}{30}$$

Again, from $\mathcal{M}_1(\mathcal{E})$ we infer that the unsolvability of the MARE is provoked by only one of its three sub-equations. However, in this case, the high value of $\mathcal{M}_2(\mathcal{E})$ means that the unsolvability of such sub-equation is strong. In fact, we can check that the sub-equation $0.3\, x_1 \vee 0.4\, x_2 = 1$ is actually unsolvable.

5 Conclusions and Future Work

A first approach to measure the solvability of a MARE has been presented. Three different forms of measuring solvability have been introduced, analysing different aspects, as the number of unsolvable sub-equations, the maximum unsolvability of the sub-equations or the average unsolvability. Additionally, some examples show the possible use of these measures to classify determine the nature of the unsolvability of a MARE.

References

1. Aliannezhadi, S., Abbasi Molai, A.: A new algorithm for geometric optimization with a single-term exponent constrained by bipolar fuzzy relation equations. Iran. J. Fuzzy Syst. **18**(1), 137–150 (2021)
2. Bělohlávek, R.: Sup-t-norm and inf-residuum are one type of relational product: unifying framework and consequences. Fuzzy Sets Syst. **197**, 45–58 (2012)
3. Cornejo, M.E., Díaz-Moreno, J.C., Medina, J.: Multi-adjoint relation equations: a decision support system for fuzzy logic. Int. J. Intell. Syst. **32**(8), 778–800 (2017)
4. Cornejo, M.E., Medina, J., Ramírez-Poussa, E.: Algebraic structure and characterization of adjoint triples. Fuzzy Sets Syst. **425**, 117–139 (2021)

5. Di Martino, F., Sessa, S.: Comparison between images via bilinear fuzzy relation equations. J. Ambient. Intell. Hum. Comput. **9**(5), 1517–1525 (2018)
6. Di Nola, A., Sanchez, E., Pedrycz, W., Sessa, S.: Fuzzy Relation Equations and Their Applications to Knowledge Engineering. Kluwer Academic Publishers, Norwell, MA, USA (1989)
7. Díaz, J.C., Medina, J.: Multi-adjoint relation equations: Definition, properties and solutions using concept lattices. Inf. Sci. **253**, 100–109 (2013)
8. Díaz-Moreno, J.C., Medina, J., Turunen, E.: Minimal solutions of general fuzzy relation equations on linear carriers. an algebraic characterization. Fuzzy Sets Syst. **311**, 112–123 (2017)
9. Dubois, D., Prade, H.: Fuzzy relation equations and causal reasoning. Fuzzy Sets Syst. **75**(2), 119–134 (1995)
10. Lobo, D., Cornejo, M.E., Medina, J.: Abductive reasoning in normal residuated logic programming via bipolar max-product fuzzy relation equations. In: 2019 Conference of the International Fuzzy Systems Association and the European Society for Fuzzy Logic and Technology (EUSFLAT 2019). Atlantis Studies in Uncertainty Modelling, vol. 1, pp. 588–594. Atlantis Press (2019)
11. Madrid, N., Ojeda-Aciego, M.: Measuring inconsistency in fuzzy answer set semantics. IEEE Trans. Fuzzy Syst. **19**(4), 605–622 (2011)
12. Medina, J.: Multi-adjoint property-oriented and object-oriented concept lattices. Inf. Sci. **190**, 95–106 (2012)
13. Medina, J.: Minimal solutions of generalized fuzzy relational equations: clarifications and corrections towards a more flexible setting. Int. J. Approx. Reason. **84**, 33–38 (2017)
14. Pedrycz, W.: Fuzzy relational equations with generalized connectives and their applications. Fuzzy Sets Syst. **10**(1–3), 185–201 (1983)

Efficiency of Fuzzy Rough Set Decision Algorithms

Fernando Chacón-Gómez, M. Eugenia Cornejo, Jesús Medina, and Eloísa Ramírez-Poussa

1 Introduction

In rough set theory (RST), decision tables can be described from decision rules [8, 15–17, 19]. Decision rules are used to extract information and to infer conclusions in an efficient way. The notion of decision algorithm was proposed by Pawlak in [16], to establish the properties that must be satisfied by a set of decision rules in order to suitably describe the decision table. An essential measure associated with decision algorithms is the notion of efficiency. The greater the efficiency of a decision algorithm is, the better this decision algorithm represents the considered dataset. As a consequence, the notion of efficiency is very useful to compare decision algorithms.

RST has already been considered within the fuzzy paradigm in different papers [2, 3, 6, 7, 9–14, 18]. Continuing the study started in [1], this work addresses the study of efficiency of decision algorithms, within the fuzzy paradigm. We provide a new notion of efficiency for fuzzy rough sets decision algorithms. In addition, we

Partially supported by the 2014–2020 ERDF Operational Programme in collaboration with the State Research Agency (AEI) in projects PID2019-108991GB-I00 and PID2022-137620NB-I00, and with the Department of Economy, Knowledge, Business and University of the Regional Government of Andalusia in project FEDER-UCA18-108612, and by the European Cooperation in Science and Technology (COST) Action CA17124.

F. Chacón-Gómez (✉) · M. E. Cornejo · J. Medina · E. Ramírez-Poussa
Department of Mathematics, University of Cádiz, Cádiz, Spain
e-mail: fernando.chacon@uca.es

M. E. Cornejo
e-mail: mariaeugenia.cornejo@uca.es

J. Medina
e-mail: jesus.medina@uca.es

E. Ramírez-Poussa
e-mail: eloisa.ramirez@uca.es

M. E. Cornejo et al. (eds.), *Computational Intelligence and Mathematics for Tackling Complex Problems 5*, Studies in Computational Intelligence 1127,
https://doi.org/10.1007/978-3-031-46979-4_3

analyze some interesting properties of this measure, proving that this novel notion of efficiency generalizes the usually considered one in the classical framework. We also include an example to illustrate the content.

2 Fuzzy Rough Set Decision Algorithms

Datasets are represented as decision tables $S = (U, \mathcal{A}_d, \mathcal{V}_{\mathcal{A}_d}, \overline{\mathcal{A}_d})$ in rough set theory, where U and $\mathcal{A}$ are non-empty sets of objects and attributes respectively, $\mathcal{A}_d = \mathcal{A} \cup \{d\}$ with $d \notin \mathcal{A}$, $\mathcal{V}_{\mathcal{A}_d} = \{V_a \mid a \in \mathcal{A}_d\}$, where V_a is the set of values associated with the attribute $a \in \mathcal{A}_d$ over U, and $\overline{\mathcal{A}_d} = \{\bar{a} \mid a \in \mathcal{A}_d, \bar{a}\colon U \to V_a\}$. The structure of the decision tables where an attribute is highlighted allows a direct transformation of them to a set of decision rules, since in both settings certain conditions must be satisfied to carry out a specific decision. Hence, the decision rules can describe the decision table in logical terms, providing other perspective of decision tables [16] and facilitating the extraction of information. To do the mentioned transformation, we need to define a formal language. In the fuzzy framework, the set of formulas associated with $B \subseteq A$, denoted as $For(B)$, is the same as in the classical framework, that is, it is built from attribute-value pairs (a, v), where $a \in B$ and $v \in V_a$, by using the conjunction and disjunction logical connectives. However, the interpretation of formulas is different.

Definition 1 Let S be a decision table and $B \subseteq \mathcal{A}$. Given $\Phi \in For(B)$, with $\Phi = (a, v)$, $a \in B$ and $v \in V_a$, the mapping $\|\Phi\|_S^E\colon U \longrightarrow [0, 1]$ is defined, for each $x \in U$, as:

$$\|\Phi\|_S^E(x) = E_a(\bar{a}(x), v)$$

where $E_a\colon V_a \times V_a \to [0, 1]$ is a $[0, 1]$-fuzzy tolerance relation, that is, a fuzzy reflexive and symmetric relation. Therefore, $\|\Phi\|_S^E(x)$ represents *how much the object x satisfies the formula Φ*, through the relationship between the value of the attribute a in the object x and the value of the attribute a in the formula Φ.

Given $\Phi, \Psi \in For(B)$, $\|\Phi \wedge \Psi\|_S^E(x)$ and $\|\Phi \vee \Psi\|_S^E(x)$ represent *how much the object xsatisfies the formula $\Phi \wedge \Psi$* and formula $\Phi \vee \Psi$, respectively, both are defined as follows:

$$\|\Phi \wedge \Psi\|_S^E(x) = \min\{\|\Phi\|_S^E(x), \|\Psi\|_S^E(x)\}$$
$$\|\Phi \vee \Psi\|_S^E(x) = \max\{\|\Phi\|_S^E(x), \|\Psi\|_S^E(x)\}$$

Decision rules are expressions $\Phi \to \Psi$, where $\Phi \in For(B), \Psi \in For(\{d\})$. In the following, we need to consider a family of $[0, 1]$-fuzzy tolerance relations, known as *separable relations* [20], in order to introduce the notions of *support* and *strength* of a decision rule in the fuzzy framework.

Definition 2 Let S be a decision table, $\Phi \to \Psi$ be a decision rule in S and $E = \{E_a : V_a \times V_a \to [0, 1] \mid a \in \mathcal{A}_d\}$ be a family of [0, 1]-fuzzy tolerance relations such that $E_a(\bar{a}(x), v) = 1$ if and only if $\bar{a}(x) = v$.

- *E-support* of the decision rule $\Phi \to \Psi$ is $supp_S^E(\Phi, \Psi) = card_F(\|\Phi \wedge \Psi\|_S^E)$ where $card_F(\cdot)$ represents the cardinal of a fuzzy set [5].
- *E-strength* of the decision rule $\Phi \to \Psi$ is $\sigma_S^E(\Phi, \Psi) = \dfrac{supp_S^E(\Phi, \Psi)}{|U|}$ where $|\cdot|$ represents the cardinal of a set.

These measures are key to study decision algorithms and their efficiency. A decision algorithm is a set of decision rules verifying certain conditions, whose formal definition in the fuzzy case is given below. First, we need to define a fuzzy tolerance relation which allows us to compare formulas.

Definition 3 Given a decision table S, we define the [0, 1]-fuzzy tolerance relation $R_{\mathrm{Fd}} : For(\mathcal{A}_d) \times For(\mathcal{A}_d) \to [0, 1]$, for all $\Phi, \Phi' \in For(\mathcal{A}_d)$, as follows:

$$R_{Fd}(\Phi, \Phi') = 1 - \max\{|\|\Phi\|_S^E(x) - \|\Phi'\|_S^E(x)| \mid x \in U\}$$

Given $\alpha \in [0, 1]$, the R_{Fd}-α-block is defined for each $\Phi \in For(\mathcal{A}_d)$ as:

$$[\Phi]_\alpha = \{\Phi' \in For(\mathcal{A}_d) \mid \alpha \leq R_{Fd}(\Phi, \Phi')\}$$

Definition 4 Let S be a decision table, $\alpha > 0$, $\alpha_1, \alpha_2, \alpha_4 \in [0, 1)$, $\alpha_3 \geq 0$ and $Dec(S) = \{\Phi_i \to \Psi_i\}_{i \in I}$ be a set of decision rules of S, where the index set is $I = \{1, \ldots, m\}$ and $m \geq 2$. We say that:

1. $Dec(S)$ is a set of $\alpha_1\alpha_2$-*pairwise mutually exclusive (independent)* decision rules, if each pair of decision rules $\Phi \to \Psi, \Phi' \to \Psi' \in Dec(S)$ satisfies that $\Phi = \Phi'$ or $\|\Phi \wedge \Phi'\|_S^E(x) \leq \alpha_1$ and $\Psi = \Psi'$ or $\|\Psi \wedge \Psi'\|_S^E(x) \leq \alpha_2$, for all $x \in U$.
2. $Dec(S)$ *covers* U, if $|\|\bigvee_{i=1}^m \Phi_i\|_S^E| = |\|\bigvee_{i=1}^m \Psi_i\|_S^E| = |U|$.
3. The decision rule $\Phi \to \Psi \in Dec(S)$ is α_3-*admissible in* S if $\alpha_3 < supp_S^E(\Phi, \Psi)$.
4. $Dec(S)$ *preserves the* α-*consistency of* S *with a degree* α_4 if the next inequality holds for all $x \in U$:

$$\left|POS_{\mathcal{A}}(x) - \bigvee_{\Phi \to \Psi \in Dec_\alpha^+(S)} \|\Phi\|_S^E(x)\right| \leq \alpha_4$$

where $\alpha \in [0, 1]$ and $Dec_\alpha^+(S) = \{\Phi \to \Psi \in Dec(S) \mid$ if for each $\Phi' \to \Psi' \in Dec(S)$ such that $\Phi' \in [\Phi]_\alpha$, then $\Psi' \in [\Psi]_\alpha\}$

The set of decision rules $Dec(S)$ satisfying the previous properties for the values $\alpha_1, \alpha_2, \alpha_3, \alpha_4$ is called $(\alpha_1, \alpha_2, \alpha_3, \alpha_4)_\alpha$-*decision algorithm in* S.

These notions were already studied in [1], where the readers can find more information, properties and illustrative examples about this research topic. In addition, it is important to emphasize that the value α must be fixed at the beginning depending on the flexibility looked for. After that, the rest of values can be computed to obtain a decision algorithm. The values α_1, α_2 set upper bounds to the degree of satisfaction of each pair of antecedents and consequents for each object of the decision table, the value α_3 provides a lower bound of the E-support of all decision rules of $Dec(S)$ and the value α_4 measures the equivalence between the consistency provided by the fuzzy positive region and the α-consistency provided by the antecedents of decision rules of $Dec^+_\alpha(S)$.

3 Efficiency of Decision Algorithms

This section addresses the study of the efficiency of fuzzy rough set decision algorithms, recalling its formal definition to the fuzzy framework and presenting different properties related to this notion.

Definition 5 Let S be a decision table and $Dec(S)$ a $(\alpha_1, \alpha_2, \alpha_3, \alpha_4)_\alpha$-decision algorithm. Given $0 \leq \varepsilon \leq 1$, we call *$\varepsilon$-efficiency* of the decision algorithm $Dec(S)$ to the number

$$\eta^\varepsilon(Dec(S)) = \sum_{\Phi \in For(\mathcal{A})} \eta^\varepsilon_\Phi(Dec(S))$$

where $\eta^\varepsilon_\Phi(Dec(S)) = \max\{\sigma^E_S(\Phi', \Psi') \mid \Phi \to \Psi, \Phi' \to \Psi' \in Dec(S), \Phi' \in [\Phi]_\varepsilon\}$.

This notion is indispensable to measure the usefulness of a decision algorithm and compare with other algorithms as well. The following proposition includes some properties associated with the notion of ε-efficiency.

Proposition 1 *Let S be a decision table and $Dec(S)$ be a $(\alpha_1, \alpha_2, \alpha_3, \alpha_4)_\alpha$-decision algorithm and $m \geq 2$. The following properties hold:*

1. *If $E = \{E_a \colon V_a \times V_a \to \{0, 1\} \mid a \in \mathcal{A}_d\}$ is a family of boolean tolerance relations and $\varepsilon > 0$ then $\eta^\varepsilon(Dec(S)) = \eta(Dec(S))$.*[1]
2. *The ε-efficiency is decreasing in ε.*
3. *If $\max\{R_{Fd}(\Phi_i, \Phi_j) \mid \Phi_i \neq \Phi_j, i, j \in \{1, 2, \ldots, m\}\} < \varepsilon$ and $\Phi_i \neq \Phi_j$ when $i \neq j$, then the ε-efficiency of the decision algorithm $Dec(S)$ is given as*

$$\eta^\varepsilon(Dec(S)) = \sum_{\Phi \to \Psi \in Dec(S)} \sigma^E_S(\Phi, \Psi)$$

4. *Let $\Phi_j \to \Psi_j$ be a decision rule in $Dec(S)$, with $j \in \{1, 2, \ldots, m\}$, such that*

[1] The notion of efficiency of a decision algorithm in the classical case, $\eta(Dec(S))$, can be found in [16].

$$\sigma_S^E(\Phi_j, \Psi_j) = \max\{\sigma_S^E(\Phi, \Psi) \mid \Phi \rightarrow \Psi \in Dec(S)\}$$

If $\varepsilon \leq \min\{R_{Fd}(\Phi_i, \Phi_j) \mid i \in \{1, 2, \ldots, m\}\}$ *then the* ε*-efficiency of the decision algorithm* $Dec(S)$ *is given as*

$$\eta^{\varepsilon}(Dec(S)) = \sigma_S^E(\Phi_j, \Psi_j) \cdot |\{\Phi \mid \Phi \rightarrow \Psi \in Dec(S)\}|$$

Next remarks are needed to emphasize the importance of these properties.

- From the first item in Proposition 1, we can conclude that Definition 5 generalizes the notion of efficiency given in the classical case. Moreover, it is convenient to mention that the condition $\varepsilon > 0$ is required because otherwise all the antecedents would be considered as equals.
- With respect to the second item in Proposition 1, the comparison of the antecedents of the decision rules in $Dec(S)$ becomes more restrictive, when the value of ε increases, since the cardinal of R_{Fd}-ε-blocks decreases. Therefore, it is expected that the ε-efficiency decreases. This property is fundamental since the user must choose a balanced value of ε to obtain a good value of ε-efficiency while a good level of flexibility is considered, which is given by ε.
- From the third item in Proposition 1, if the fixed value ε allows to discern all antecedents of decision rules in $Dec(S)$, then only one decision rule is considered in the computation of each addend of the ε-efficiency. For this reason and the previous property, this ε-efficiency is the minimum possible value for the decision algorithm $Dec(S)$.
- The last item states that when the antecedent of the decision rule with the greatest value of E-strength is R_{Fd}-ε-related with the rest of antecedents, this decision rule is considered in each addend of the ε-efficiency. As a consequence, the computation of ε-efficiency boils down to the product of E-strength times the number of different antecedents in $Dec(S)$. In addition, under the hypothesis required in the fourth item, we can ensure the decision rule $\Phi_j \rightarrow \Psi_j$ is the most representative of the table. Therefore, this ε-efficiency is the maximum possible value for $Dec(S)$.

The following example continues the one given in [1] in order to illustrate Definition 5 and Proposition 1.

Example 1 Consider the decision table $S = (U, \mathcal{A}_d, \mathcal{V}_{\mathcal{A}_d}, \overline{\mathcal{A}_d})$ where the set of objects is $U = \{x_1, x_2, x_3, x_4, x_5, x_6, x_7\}$, the set of attributes is $\mathcal{A} = \{a_1, a_2, a_3\}$ and $\alpha = 0.75$. This decision table is represented in Table 1, together with the E-Support and E-strength of the decision rules given by each object and the relation between each pair of antecedents of decision rules. The considered decision rules can be found in [1] and they are easily obtained from the row of S.

Taking $\alpha_1 \geq 0.78$, $\alpha_2 \geq 0$, $\alpha_3 < 1.61$ and $\alpha_4 \geq 0.83$, we obtain that $Dec(S)$ is a $(\alpha_1, \alpha_2, \alpha_3, \alpha_4)_{0.75}$-decision algorithm. Now, we compute the ε-efficiency of $Dec(S)$ by using Definition 5, Proposition 1 and considering different values of ε.

Table 1 Decision table, E-support, E-strength and relation R_{Fd} of Example 1

S	a_1	a_2	a_3	d
x_1	0.34	0.31	0.75	0
x_2	0.21	0.71	0.5	1
x_3	0.52	0.92	1	0
x_4	0.85	0.65	1	1
x_5	0.43	0.89	0.5	0
x_6	0.21	0.47	0.25	1
x_7	0.09	0.93	0.25	0

Rules	$supp_S^E$	σ_S^E
r_1	2.19	0.31
r_2	2.11	0.3
r_3	2.14	0.31
r_4	1.61	0.23
r_5	2.58	0.37
r_6	2	0.29
r_7	2.29	0.33

R_{Fd}	Φ_1	Φ_2	Φ_3	Φ_4	Φ_5	Φ_6	Φ_7
Φ_1	1	0.6	0.39	0.49	0.42	0.5	0.38
Φ_2	0.6	1	0.5	0.36	0.78	0.75	0.75
Φ_3	0.39	0.5	1	0.67	0.5	0.25	0.25
Φ_4	0.49	0.36	0.67	1	0.5	0.25	0.24
Φ_5	0.42	0.78	0.5	0.5	1	0.58	0.66
Φ_6	0.5	0.75	0.25	0.25	0.58	1	0.54
Φ_7	0.38	0.75	0.25	0.24	0.66	0.54	1

- To begin with, we will compute the 0.75-efficiency of $Dec(S)$ by using Definition 5. From the relation R_{Fd} given in Table 1 it is easy to obtain the R_{Fd}-0.75-blocks of the antecedents:

$$[\Phi_1]_{0.75} = \{\Phi_1\}$$
$$[\Phi_2]_{0.75} = \{\Phi_2, \Phi_5, \Phi_6, \Phi_7\} \quad [\Phi_4]_{0.75} = \{\Phi_4\} \qquad [\Phi_6]_{0.75} = \{\Phi_2, \Phi_6\}$$
$$[\Phi_3]_{0.75} = \{\Phi_3\} \qquad [\Phi_5]_{0.75} = \{\Phi_2, \Phi_5\} \quad [\Phi_7]_{0.75} = \{\Phi_2, \Phi_7\}$$

Moreover, the decision rules with the antecedents whose R_{Fd}-0.75-blocks are not singleton and ranked in order of E-strength from the highest to the lowest are r_5, r_7, r_2 and r_6. Therefore, the 0.75-efficiency of $Dec(S)$ is

$$\begin{aligned}\eta^{0.75}(Dec(S)) &= \sigma_S^E(\Phi_1, \Psi_1) + \max\{\sigma_S^E(\Phi_2, \Psi_2), \sigma_S^E(\Phi_5, \Psi_5), \sigma_S^E(\Phi_6, \Psi_6), \sigma_S^E(\Phi_7, \Psi_7)\}\\ &\quad + \sigma_S^E(\Phi_3, \Psi_3) + \sigma_S^E(\Phi_4, \Psi_4) + \max\{\sigma_S^E(\Phi_2, \Psi_2), \sigma_S^E(\Phi_5, \Psi_5)\}\\ &\quad + \max\{\sigma_S^E(\Phi_2, \Psi_2), \sigma_S^E(\Phi_6, \Psi_6)\} + \max\{\sigma_S^E(\Phi_2, \Psi_2), \sigma_S^E(\Phi_7, \Psi_7)\}\\ &= \sigma_S^E(\Phi_1, \Psi_1) + \sigma_S^E(\Phi_5, \Psi_5) + \sigma_S^E(\Phi_3, \Psi_3) + \sigma_S^E(\Phi_4, \Psi_4) +\\ &\quad + \sigma_S^E(\Phi_5, \Psi_5) + \sigma_S^E(\Phi_2, \Psi_2) + \sigma_S^E(\Phi_7, \Psi_7)\\ &= 2.22\end{aligned}$$

- To exemplify the third item in Proposition 1, we observe that there are no decision rules with the same antecedent (see [1]), and by using the relation R_{Fd} given in Table 1, we can take:

$$\varepsilon > \max\{R_{Fd}(\Phi_i, \Phi_j) \mid \Phi_i \neq \Phi_j, i, j \in \{1, 2, \ldots, 7\}\} = 0.78$$

in order to apply Proposition 1, obtaining that the ε-efficiency of $Dec(S)$ is:

$$\eta^{\varepsilon}(Dec(S)) = \sum_{\Phi \to \Psi \in Dec(S)} \sigma_S^E(\Phi, \Psi) = 2.14$$

- Finally, we illustrate the fourth item in Proposition 1. From Table 1, the decision rule in $Dec(S)$ with the greatest value of E-strength is r_5, that is, $\sigma_S^E(\Phi_5, \Psi_5) = 0.37$. On the other hand, from Table 1 again, we have that:

$$\min\{R_{Fd}(\Phi_i, \Phi_5) \mid i \in \{1, 2, \ldots, 7\}\} = R_{Fd}(\Phi_1, \Phi_5) = 0.42$$

As a consequence, if $\varepsilon \leq 0.42$, by using Proposition 1, we obtain that

$$\begin{aligned}\eta^{\varepsilon}(Dec(S)) &= \max\{\sigma_S^E(\Phi, \Psi) \mid \Phi \to \Psi \in Dec(S)\}|\{\Phi \mid \Phi \to \Psi \in Dec(S)\}| \\ &= 0.37 \cdot 7 = 2.59\end{aligned}$$ □

Attending to these computations, we obtain that the minimum and maximum value of ε-efficiency are similar. Hence, the fact of carrying out a less flexible study of the ε-efficiency of the decision algorithm $Dec(S)$, that is, considering a higher value of ε, does not have a great effect on the obtained results. This aspect is quite important since it will allow to discern decision rules while maintaining the ε-efficiency of $Dec(S)$ without significant changes.

4 Conclusions and Further Work

In this paper, we have provided a new notion of efficiency, which can be used to measure how well a fuzzy rough set decision algorithm describes a dataset. We have proved that this notion generalizes the classical one and we have presented some interesting properties. We also have included an example to illustrate the generalized notion and its properties.

As future work, we want to continue studying more properties about decision algorithms in the fuzzy environment, as well as applying all the obtained results in real datasets. Taking into account the existing connection between the given notions of E-support and E-strength and those of support and confidence from fuzzy association rule mining, we will consider the alternative measures introduced in [4] and investigate how they could impact in our work.

References

1. Chacón-Gómez, F., Cornejo, M.E., Medina, J., Ramírez-Poussa, E.: Fuzzy rough set decision algorithms. Commun. Comput. Inf. Sci. **1601**, 63–76 (2022)
2. Cock, M.D., Cornelis, C., Kerre, E.E.: Fuzzy rough sets: the forgotten step. IEEE Trans. Fuzzy Syst. **15**(1), 121–130 (2007)
3. Cornelis, C., Medina, J., Verbiest, N.: Multi-adjoint fuzzy rough sets: definition, properties and attribute selection. Int. J. Approx. Reason. **55**, 412–426 (2014)
4. De Cock, M., Cornelis, C., Kerre, E.: Elicitation of fuzzy association rules from positive and negative examples. Fuzzy Sets Syst. **149**(1), 73–85 (2005)
5. De Luca, A., Termini, S.: A definition of a nonprobabilistic entropy in the setting of fuzzy sets theory. Inf. Control **20**(4), 301–312 (1972)
6. Dubois, D., Prad, H.: Rough fuzzy sets and fuzzy rough sets. Int. J. Gen. Syst. **17**(2–3), 191–209 (1990)
7. Feng, T., Fan, H.-T., Mi, J.-S.: Uncertainty and reduction of variable precision multigranulation fuzzy rough sets based on three-way decisions. Int. J. Approx. Reason. **85**, 36–58 (2017)
8. Li, J., Mei, C., Lv, Y.: Incomplete decision contexts: Approximate concept construction, rule acquisition and knowledge reduction. Int. J. Approx. Reason. **54**(1), 149–165 (2013)
9. Medina, J.: Towards multi-adjoint property-oriented concept lattices. Lect. Notes Artif. Intell. **6401**, 159–166 (2010)
10. Medina, J.: Multi-adjoint property-oriented and object-oriented concept lattices. Inf. Sci. **190**, 95–106 (2012)
11. Medina, J.: Relating attribute reduction in formal, object-oriented and property-oriented concept lattices. Comput. Math. Appl. **64**(6), 1992–2002 (2012)
12. Morsi, N.N., Yakout, M.M.: Axiomatics for fuzzy rough sets. Fuzzy Sets Syst. **100**, 327–342 (1998)
13. Nakamura, A.: Fuzzy rough sets. Note Mult.-Valued Log. Jpn. **9**, 1–8 (1988)
14. Nanda, S., Majumdar, S.: Fuzzy rough sets. Fuzzy Sets Syst. **45**, 157–160 (1992)
15. Pawlak, Z.: Rough Sets: Theoretical Aspects of Reasoning About Data. Kluwer Academic Publishers, Norwell, MA, USA (1992)
16. Pawlak, Z.: Rough sets and decision algorithms. In: Ziarko, W., Yaopages, Y. (eds.) Rough Sets and Current Trends in Computing, pp. 30–45. Springer, Berlin (2001)
17. Stawicki, S., Ślęzak, D., Janusz, A., Widz, S.: Decision bireducts and decision reducts—a comparison. Int. J. Approx. Reason. **84**, 75–109 (2017)
18. Wang, C.Y.: Topological structures of l-fuzzy rough sets and similarity sets of l-fuzzy relations. Int. J. Approx. Reason. **83**, 160–175 (2017)
19. Yao, Y.: Three-way decisions with probabilistic rough sets. Inf. Sci. **180**(3), 341–353 (2010)
20. Zedam, L., Bouremel, H., De Baets, B.: Left- and right-compatibility of order relations and fuzzy tolerance relations. Fuzzy Sets Syst. **360**, 65–81 (2019)

Use of Fuzzy Time Series to Generate Linguistic Descriptions of Noise Pollution

Luis Rodriguez-Benitez, Juan Moreno-Garcia, Ester del Castillo-Herrera, Jun Liu, and Luis Jimenez-Linares

1 Introduction

A Time Series (TS) can be defined as a set of time-ordered regular observations of a quantitative characteristic of a phenomenon measured at consecutive points in time. It is possible to use fuzzy logic for modelling time series and this allows to approximate functions incorporating the readability associated to linguistic variables. One option to model these series is the use of Fuzzy Time Series (FTS) which are composed of fuzzy sets [1, 2]. Their foundations were defined by Song and Chissom in [3]. In this work we propose a way of representing sets of TS by means of FTS composed of linguistic labels previously defined in a set. The sets of TS taken as input are represented by a FTS, called *initial FTS*, following the method presented in [4]. In order to obtain a more understandable representation, the *initial FTS* is transformed into a FTS composed of linguistic labels that have been selected from a predefined set with labels defined in the domain of application. This set is called *labels FTS*.

L. Rodriguez-Benitez (✉) · E. Castillo-Herrera · L. Jimenez-Linares
Escuela Superior de Informatica, Universidad de Castilla-La Mancha, Ciudad Real, Spain
e-mail: luis.rodriguez@uclm.es

E. Castillo-Herrera
e-mail: ester.castillo@uclm.es

L. Jimenez-Linares
e-mail: luis.jimenez@uclm.es

J. Moreno-Garcia
Escuela de Ingenieria Industrial y Aeroespacial, Universidad de Castilla-La Mancha, Toledo, Spain
e-mail: juan.moreno@uclm.es

J. Liu
School of Computing, University of Ulster, Belfast, Northern Ireland, UK
e-mail: j.liu@ulster.ac.uk

M. E. Cornejo et al. (eds.), *Computational Intelligence and Mathematics for Tackling Complex Problems 5*, Studies in Computational Intelligence 1127,
https://doi.org/10.1007/978-3-031-46979-4_4

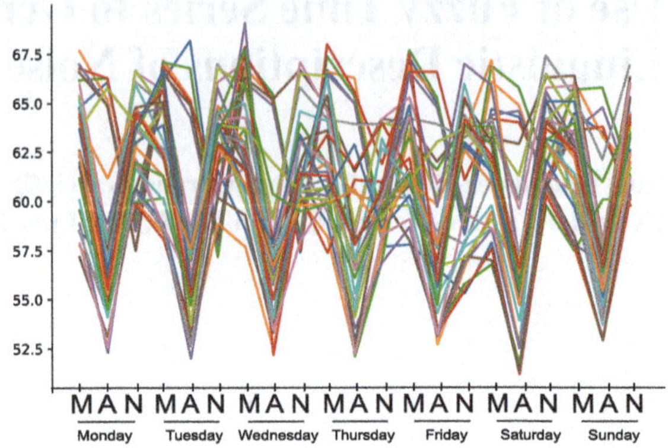

Fig. 1 TS of daily L_{Aeq} measurement in the morning, afternoon and night

The Monte Carlo Method [5] has been used to perform this transformation. As case use, we apply this method to generate linguistic descriptions of noise pollution data taken from environmental stations. Noise measurement stations typically use the A-weighted sound pressure level (dBA) to measure noise in an area as described in [6]. To complete the tests of this work, noise data have been extracted from the website of the "Portal de datos abiertos del Ayuntamiento de Madrid".[1] For each day, four periods are distinguished, providing six values for each period. The first period corresponds to the whole day (T) while the other three periods divide the day into the day period (M) from 7:00 to 19:00, the afternoon period (A) from 19:00 to 23:00 and the night period (N) from 23:00 to 7:00. The A-weighted equivalent continuous sound pressure level determined over the whole day is the measure L_{Aeq}. The measurements L_{AS01}, L_{AS10}, L_{AS50}, L_{AS90} and L_{AS99} show the noise exceeded during 1%, 10%, 50%, 90% and 99% respectively. The values of L_{AS01} and L_{AS99} are close to the minimum and maximum noise reached during the studied period, respectively. Figure 1 shows the set of 33 series studied, showing data of noise for the first 33 weeks of the year 2020 monitored by the Cuatro Caminos station in Madrid (Spain). The rest of the paper is organised as follows: Sect. 2 presents the proposed method and details its testing for the generation of noise pollution descriptions. Finally, Sect. 3 presents the conclusions and future work.

2 Description of the Proposal

This section presents the generation of the *initial FTS* and the generation of the new FTS using the linguistic labels from the predefined set.

[1] https://cutt.ly/Lj6O0xP.

2.1 Generation of the Original FTS

A method that aggregates a set of TS into a FTS is used to obtain the *initial FTS* [4]. This method uses a collection of TS as input. A FTS $F(t)$ is formalised as a collection of fuzzy sets $f_i(t)$ over the universe of discourse $Y(t)$, where $Y(t)$ is a subset of $\mathbb{R}$, $t = (0, 1, 2, \ldots)$. It is noted that the observations of a FTS are fuzzy sets while those of a TS consist of real numbers [3].

A collection of TS is represented as $S = \{s_0, \ldots, s_{n-1}\}$, where each TS is formalised as $s_i = \{v_{(i,t_0)}, v_{(i,t_1)}, \ldots, v_{(i,t_{m-1})}\}$ where t_j represents the time of each observation. Let $Y(t) = \{v_{(0,t)}, \ldots, v_{(n-1,t)}\}$ with $t \in (t_0, \ldots, t_{m-1})$ and $DE_t(v)$ a density function that estimates the probability of v belonging to $Y(t)$. The fuzzy set f_t over the domain $Y(t)$ is defined as the membership function $\mu_t(v)$ (Eq. 1).

$$\mu_t(v) = \frac{DE_t(v)}{max\{DE_t(v)|v \in Y(t)\}} \tag{1}$$

In [4], the KDE (Kernel Density Estimation) technique is used [7, 8] with Gaussian kernel functions to obtain the DE_t density functions. The FTS can now be defined on the collection S: $FTS(S) = \{f_{t_0}, f_{t_1}, \ldots, f_{t_{m-1}}\}$. This definition is extended to an arbitrary set T with $t \subset T$ such that if $r \in T$, its membership function f_r is shown in Eq. 2.

$$\mu_r(v) = \begin{cases} \frac{DE_r(v)}{max\{DE_r(v)|v \in Y(r)\}} & r \in t \\ \mu_{t_{down}}(v)(1-\alpha) + \alpha\mu_{t_{up}}(v) & t_0 \leq r \leq t_{m-1}, r \notin t \\ 0 & r < t_0 \ or \ r > t_m \end{cases} \tag{2}$$

being $t_{down} = max_j\{j < r | j \in t\}$, $t_{up} = min_i\{i > r | i \in t\}$ and $\alpha = \frac{r - t_{down}}{t_{up} - t_{down}}$.

The result of calculating the FTS of the set of series shown in Fig. 1 is presented in Fig. 2 where for each time instant t of the FTS, the membership function $\mu_t(v)$ is represented as a vertical line where its thickness and colour indicate its value. For example, it can be observed that on Monday morning the membership function reflecting L_{Aeq} has its maximum value between 62.5 and 65 as indicated by its blue colour and its greater thickness.

2.2 Matching Linguistic Labels and Fuzzy Sets

This section shows how to calculate the correspondence between the membership functions of *TFS* with the labels of a predefined set. Consider a linguistic variable V defined on the same domain as $\mu_t(x)$, with a set of r labels $L = \{e_0, \ldots, e_{r-1}\}$ associated with a semantics by means of a membership function $\mu_{e_i}(x)$ for $\{i = 0, \ldots, r-1\}$. In the use case, the linguistic variable L_{Aeq}, is defined in the interval

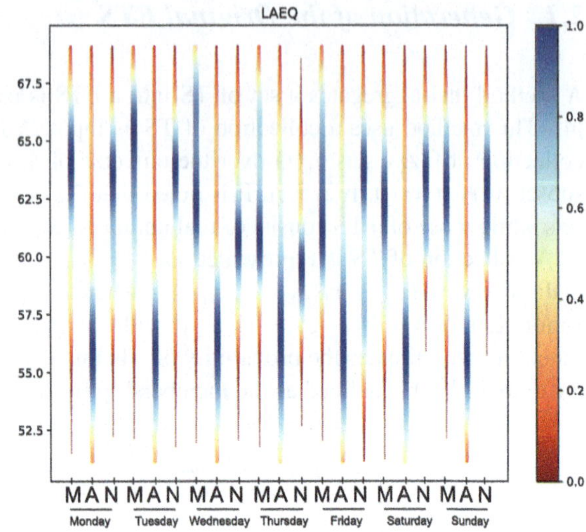

Fig. 2 FTS resulting from the aggregation of 33 weekly TS using the measure L_{Aeq} for all days of the week into morning, afternoon and evening

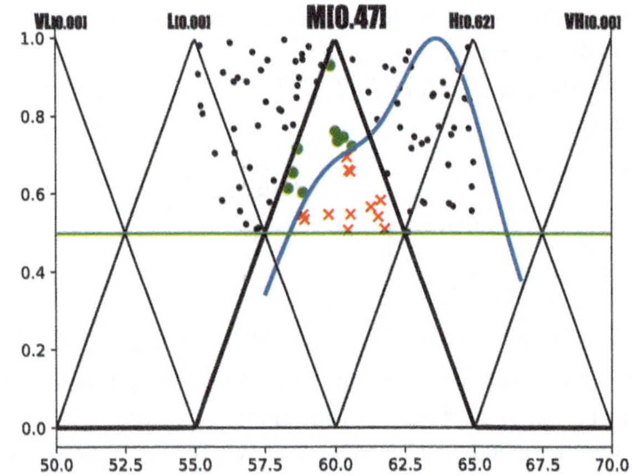

Fig. 3 Set of random points generated in the Montecarlo estimation of $o_{t,i}$

[50dB,70dB] and has a set of 5 labels: Very Low, Low, Medium, High, Very High ([VL,L,M.H,VH]) with triangular membership functions as shown in Fig. 3.

To perform the matching we estimate $o_{t,i}$, the overlap percentage between the membership function of the FTS $\mu_t(x)$ at a specific time instant t and the labels with membership $\mu_{e_i}(x)$ of the variable V. It is used an approximation similar to the Monte Carlo calculation integral [5] for calculating each $o_{t,i}$ (Algorithm 1). A uniformly distributed random variable with $npoint$ points is generated in the interval $[50dB, 70dB]x[\alpha, 1]$. The variable $npLabel$ stores those belonging to the area of the label and with membership values greater than α (confidence threshold value). The variable $npFTS$ contains those belonging to the area of the FTS membership function at instant t (Fig. 3). As a result, we obtain the approximation $o_{t,i} = npFTS/npLabel$.

Data: $\mu_t(x)$, $\mu_{e_i}(x)$, α,npoints
Result: $o_{t,i}$

```
npLabel ⟵ 0;
npFTS ⟵ 0;
while npoints > 0 do
    // Get random point (x,y);
    x ⟵ U(D_ei) // Uniform Random in e_i domain;
    y ⟵ U([α, 1.0])// Uniform Random in [α, 1.0] domain;
    // (x,y) point in e_i area;
    if y ≤ μ_ei(x) then
        npLabel ⟵ npLabel + 1;
        // (x,y) in the areas shared by μ_ei and μ_t;
        if y ≤ μ_t(x) then
            npFTS ⟵ npFTS + 1;
        end
    end
end
o_t,i = npFST / npLabel;
```

Algorithm 1: Montecarlo $o_{t,i}$ estimation

For each time t, all the values $o_{t,i}$ of each label e_i greater than 0 are obtained and sorted in descending order, assigning the conjunction of the labels with a value $o_{t,i}$ greater than 0.2. The final result of the linguistic FTS can be seen in Table 1.

Table 1 FTS labels with $o_{i,t} > 0$

t	FTS Label	$o_{t,i}$			
Mon M	$H \longleftarrow$	(H:0.921)	(M:0.176)		
Mon A	$L \longleftarrow$	(L:0.952)	(M:0.147)		
Mon N	$H \vee M \longleftarrow$	(H:0.615)	(M:0.472)		
Tur M	$H \longleftarrow$	(H:1.000)	(M:0.326)	(VH:0.013)	
Tur A	$L \longleftarrow$	(L:0.974)	(M:0.134)	(VL:0.003)	
Tur N	$H \longleftarrow$	(H:0.616)	(M:0.102)		
Wen M	$H \vee M \longleftarrow$	(H:0.910)	(M:0.416)	(VH:0.117)	
Wen A	$L \vee M \longleftarrow$	(L:0.913)	(M:0.247)		
Wen N	$M \longleftarrow$	(M:0.822)	(H:0.037)		
Thu M	$M \vee H \longleftarrow$	(M:0.795)	(H:0.576)		
Thu A	$H \vee M \longleftarrow$	(L:0.967)	(M:0.675)	(VL:0.044)	(H:0.044)
Thu N	$M \longleftarrow$	(M:0.840)	(L:0.005)		

(continued)

Table 1 (continued)

t	FTS Label	$o_{t,i}$			
Fri M	$H \vee M \longleftarrow$	(H:0.832)	(M:0.774)	(VH:0.002)	
Fri A	$L \vee M \longleftarrow$	(L:0.989)	(M:0.575)	(H:0.025)	
Fri N	$H \vee M \longleftarrow$	(H:0.654)	(M:0.493)	(L:0.151)	
Sat M	$M \vee H \longleftarrow$	(M:0.665)	(H:0.403)		
Sat A	$L \longleftarrow$	(L:0.986)	(M:0.127)	(VL:0.061)	
Sat N	$H \vee M \longleftarrow$	(H:0.573)	(M:0.335)		
Sun M	$H \vee M \longleftarrow$	(H:0.617)	(M:0.389)		
Sun A	$L \longleftarrow$	(L:0.913)	(M:0.059)		
Sun N	$H \vee M \longleftarrow$	(H:0.579)	(M:0.402)		

3 Conclusions and Future Works

In this paper we have presented a proposal to represent sets of TS by means of FTS. The main novelty is that they are composed of linguistic labels taken from a predefined set of linguistic instead of fuzzy sets as usual in FTS. This representation offers a higher interpretability of the FTS by using linguistic labels that can be defined by an expert in the domain. To match each fuzzy set of the initial FTS with the linguistic labels defined in the set, the Monte Carlo method has been used. The presented proposal has been used for the generation of linguistic descriptions of noise pollution. As future work, we think it could be useful to work with a more advanced aggregation technique to obtain more powerful linguistic descriptions.

Acknowledgements We are grateful for support by Grants TRA2016-76914-C3-2-P and PID2020-112967GB-C32, PID2020-112967GB-C33 funded by MCIN /AEI/10.13039/501100011033 and by *ERDF A way of making Europe*.

References

1. Zadeh, L.A.: Fuzzy sets. Inf. Control **8**(3), 338–353 (1965)
2. Zadeh, L.A.: The concept of a linguistic variable and its application to approximate reasoning—I. Inf. Sci. **8**(3) (1975)
3. Song, Q., Chissom, B.S.: Fuzzy time series and its models. Fuzzy Sets Syst. **54**(3) (1993)
4. Rodriguez-Benitez, L., Moreno-Garcia, J., Castillo-Herrera, E., Liu, J., Jimenez-Linares, L.: Aggregation and definition of an algebraic framework over fuzzy time series: an application in the supply-demand domain. Int. J. Approx. Reason. **149**, 104–115 (2022)
5. Robert, C.P., Casella, G.: Montecarlo integration. In: Monte Carlo Statistical Methods, pp. 71–138. Springer, New York (1999)
6. Moreno-Garcia, J., Jimenez-Linares, L., Liu, J., Rodriguez-Benitez, J.: Generation of linguistic descriptions for daily noise pollution in urban areas. In: 2021 IEEE International Conference on Fuzzy Systems (FUZZ-IEEE), pp. 1–6 (2021)

7. Parzen, E.: On estimation of a probability density function and mode. Ann. Math. Stat. **33**(3) (1962)
8. Kristan, M., Leonardis, A., Skočaj, D.: Multivariate online kernel density estimation with Gaussian kernels. Pattern Recognit. **44**(10–11) (2011)

A Novel Fingerprint Identification Fuzzy System Using a Center-Distance Weighted Local Binary Pattern

Ahmad A. Momani and László T. Kóczy

1 Introduction

Identity verification and authentication are critical processes that keep data safe. The most used traditional ways to protect data via passwords, personal identification numbers (PINs), and cards are not secure enough. Such traditional mechanisms could be hacked, forgotten, or stolen.

To overcome the problems in traditional security solutions, the world turned towards identifying people by biological traits. The field dealing with identification based on biological and behavioral traits is called *biometric recognition*, in short, biometrics. Biometric technologies like fingerprints, face and voice recognition, ensure that only authorized people could access the data. Biometrics research is constantly working on achieving ever better results [1–3].

The most common biometric element is the fingerprint, sufficient to distinguish and recognize people due to its uniqueness. Researchers have shown that fingerprints are reliable to identify people even if they are close relatives or identical twins.

Overall, people are always concerned about the safety of their sensitive data, government, and companies as well. Thus, the development of better security mechanisms is essential to keep data safe [2].

Despite the rapid development of fingerprint identification systems, some challenging points need more enhancement. In this research, we propose a new Fuzzy

A. A. Momani (✉) · L. T. Kóczy
Department of Information Technology, Széchenyi István University, Győr, Hungary
e-mail: momani.ahmad@sze.hu

L. T. Kóczy
e-mail: koczy@sze.hu

L. T. Kóczy
Department of Telecommunications and Artificial Intelligence, Budapest University of Technology, Budapest, Hungary

M. E. Cornejo et al. (eds.), *Computational Intelligence and Mathematics for Tackling Complex Problems 5*, Studies in Computational Intelligence 1127,
https://doi.org/10.1007/978-3-031-46979-4_5

Fingerprint Identification System (FFIS), where we will draw attention to improving fingerprint matching results using Local Binary Pattern (LBP). We will present two new methods to solve the drawbacks of fingerprint rotation. We will also address the vagueness of the identification results by creating a new fuzzy approach to interpret the results by identifying the degrees of likelihood of the uncertainty. So, rather than working on the similarity of fingerprints to create classifiers, we are working at a deeper level to help rank the fingerprints properly so the investigators in forensic sciences may come up with the most accurate decision in less time. The grade of improvement and the threshold selection will depend on the value for the performance metric Equal Error Rate (EER). EER represents the equality of two values; the False Acceptance Rate (FAR) and the False Rejection Rate (FRR). FAR occurs when the system accepts unauthorized people, while FRR occurs when the system rejects authorized people.

In summary, the contributions presented in this work are ordered according to their use in the proposed FFIS as follows:

- We propose a new *rotation solver method* that can solve the rotation angle of the fingerprint without the need to repeat the matching process multiple times with different angles. The novel rotation solver method has improved the results significantly as shown in the results part of this work.
- We propose a new method to detect the *core-area* of a fingerprint instead of a single core-point. Namely, as we may process a snipped part of the image, it is not always recommended to do the process at the core-point of the fingerprint because of potential major cuts or shifts of the core-point in the image.
- We propose a new *matching method* based on different weights assigned according to the distance from the center of the snipped image. The maximum weight is assigned to the inner level as it would be closer to the part of the image where the big differences between fingerprints can be detected. Here, the weight decreases when moving towards the outer levels. Three levels will be suggested in this work.
- Finally, we propose to utilize a fuzzy approach in our system to provide a wider range of possibilities for the answer, so that the new system provides a set of candidates ranked according to the similarity with the reference fingerprint.

The paper is organized as follows: Sect. 2 presents the proposed FFIS system and describes its stages, from fingerprint image enhancement to the matching process. Section 3 shows the test results conducted using the MATLAB software program. And finally, we conclude this paper based on our system results in Sect. 4.

2 Related Works

Sánchez et al. [4] proposed two fuzzy inference systems to find the optimal parameters for their optimized neural network human recognition system. The proposed method uses the particle swarm optimization (PSO) technique to module a granular neural network applied to three databases: ear, iris, and face databases. The method has a pre-processing stage where the image is cut and resized for the first two databases and then divided into three areas of interest for all. After that, the learning and testing phase is performed with dynamic training data optimized by PSO. The results show that PSO with fuzzy measurements performed better in most cases with 50 and 80 percent training data except for the ear database with 50 percent training data.

Melin and Sánchez [5] proposed a type-2 fuzzy inference system that uses a genetic algorithm to classify Diabetes. The tests were performed on PIMA Indian Diabetes dataset which has 336 instances divided into design and testing phases with 70% and 30% respectively. The proposed algorithm used the trapezoidal membership function instead of the triangular functions designed in the type-1 fuzzy inference system. The results of the type-1 and type-2 inference systems showed the new method's effectiveness over the previous one.

3 Fuzzy Fingerprint Identification System (FFIS) Overview

The proposed fingerprint identification system FFIS was implemented in MATLAB using MATLAB matrix-based language [6]. The proposed approach is discussed as follows.

3.1 Fingerprint Enhancement and Binarization

Since the fingerprint image could be noisy, it is better to enhance the clarity of the image before processing it. In this research, the Fourier domain analysis is used to enhance the quality of the image proposed by Chikkerur et al. [7]. Then the fingerprint image is converted to binary codes (0, 1).

3.2 Rotation Solver of the Enhanced Images

Traditionally, the enhanced image should be rotated multiple times when using the examined database. Both FVC 2002 and FVC 2004 databases contain fingerprint images with rotation up to 30 degrees. To get a better matching score, the identification system should investigate the fingerprint pairs multiple times with different rotation degrees to achieve the best matching score. Example, in [8] the complete matching process was repeated seven times with rotation degrees between −30 to 30. So, to check if the fingerprint is identified in the database FVC 2002 DB1_B, you need to perform the identification process 80 times for a certain degree, a total of 560 times (80 × 7 times) just for one identification test. Besides the vast time consuming, complex computations, the rotation results in an image with a different

Fig. 1 Results of applying the rotation solver method on a fingerprint from the FVC 2002 DB1_B database

resolution for each degree, which increases the complexity of computations when processing the image before matching.

Considering the challenges of using rotation in fingerprint identification, we propose a new method as this paper's first contribution which can solve the rotation problems in just one single process in an intelligent way. The proposed rotation solver method works by shifting the center of each row of pixels to the center of the image itself. The center of each row is different, based on where the 1's start and where they end. This way a somewhat compressed, properly rotated image will be obtained when applied to fingerprints with a rotation degree less than 30 degrees. Further studies are needed to investigate the effectiveness of the rotation solver method as it is expected not to have good results with higher rotation angles, especially when approaching 90 degrees. Overall, applying the new rotation solver method has advantages:

- Time-saving: when using the new rotation solver method, one single identification process is needed to overcome the rotation problem for any rotation angle.
- Fewer complex computations are needed for further processing.
- The resolution of the resulting image is the same as the original one no matter the rotation angle.

The experimental results show significant improvement in the matching when the rotation solver method is applied. The improvements are presented in the Results section of this paper. Figure 1 shows the result of applying the new rotation solver problem to a rotated fingerprint in the FVC 2002 DB1_B database.

3.3 *Decreasing the Image Size*

Working on the center part of the image is more efficient because it usually holds the fingerprint core-point fixed. Although we have a core-point detection code, it is not optimal for use in our method, because we consider only a 200 × 200 pixels part of the image, while the image could have cuts and core point shifts to the sides of the image (see Table 1). Such cases may have serious consequences affecting the matching results due to partial, empty spaces, or even, the matching area may be

Table 1 Processing selected area of 200 × 200 pixels: core-point and core-area

Method	Example of comparing core-area and core-point
Core-point	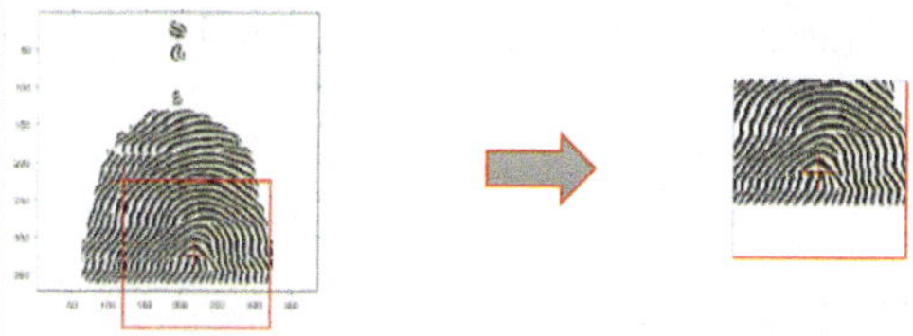
Core-area	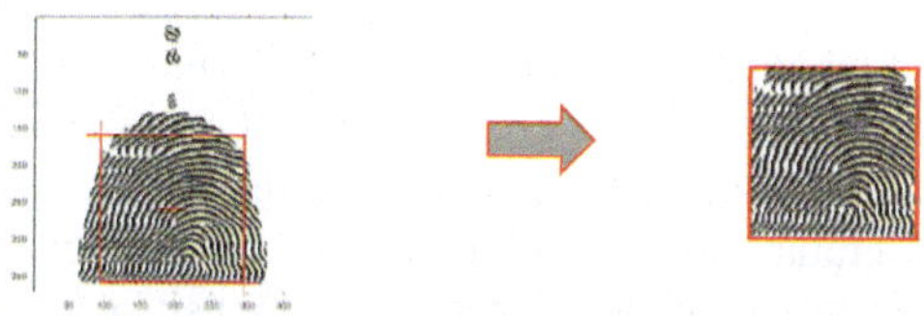

decreased significantly. To overcome this challenge, we propose to detect the core-area instead of detecting just the core-point of the fingerprint and thus utilizing the richest part of the fingerprint. After that, snipping the fingerprint image to a smaller size will improve the fingerprint matching speed and accuracy. Processing the whole fingerprint with a lot of image distortion and noise would decrease the total accuracy indeed.

3.4 Local Binary Pattern Weighted Vector Extraction

In this step, we use the local binary pattern to calculate the feature vectors for the snipped sub-window around the core point. Then, we calculate the distance between the original input image and the database images to find differences between the images. Next, we divide the 200 × 200 pixels image into sub-windows of 40 × 40 pixel size. Finally, the distance extracted will be multiplied by the block membership degrees. Here, we have three different weights based on the distance from the center (0, 0).

3.5 Matching

The next is to decide what is the suitable linguistic interpretation of the total distance of the two fingerprint images (input and database) calculated, combined with fuzzy linguistic interpretations of the range between the minimum and maximum degrees of trust (similarity < 97.6%, and > 98.4473%, respectively). We also assign a rank of similarity based on the comparison between the result with a fingerprint and the average result for each person in the database. FVC 2002 has fingerprints of 10 persons with 8 impressions for each. The result for the examined database will be a set of 3 candidates ranked based on the highest similarity to fingerprints in the database.

4 Results and Discussion

The proposed FFIS aims to compare the input fingerprint to the database and then if not rejected, the new system will give the closest three candidates for further investigation. This helps detect if the person is in the database or not, better than all methods of the literature, as the traditional methods cannot find an FRR value equal to zero. FFIS is substantial when working with sensitive fields such as forensic sciences which cannot rely on automatic fingerprint recognition only. We applied our method to all fingerprints in the FVC2002 DB1-B database. The results are shown in the following table.

Here, the system will not provide candidates with the "Certainly No" answer. While for other answers, the system will provide three candidates ordered based on their similarity to the input fingerprint.

As it is shown in Tables 2 and 3, the new approach succeeded to identify all fingerprints stating that they all belong to the database with various certainty levels. So, the approach provides three candidates for each matching process. In Table 3, for a "Certainly Yes" answer all matching cases (64 cases) have been classified correctly, except for the two cases shown in the Table. For "Most probably Yes" and "Possibly Yes" all cases are shown.

Table 2 List of possible answers when comparing the fingerprints of the FVC2002 database with itself

Database				Number of fingerprints		
FVC 2002 DB1-B				80		
Possible answers						
Certainly yes	Most probably yes	Possibly yes	Cannot say	Possibly no	Most probably no	Certainly no
64	13	3	0	0	0	0

Table 3 Shows the candidates and the correct answer

	Fingerprint compared to DB	Candidate 1	Candidate 2	Candidate 3	Correct answer
Certainly yes	104_5	108	104	109	104
	110_4	107	110	105	110
Most probably yes	101_5	101	102	105	101
	101_6	101	102	103	101
	103_1	103	109	105	103
	103_2	103	101	110	103
	103_4	103	110	106	103
	103_5	103	109	105	103
	103_6	103	105	102	103
	103_8	103	109	105	103
	104_4	104	109	102	104
	107_6	107	110	102	107
	110_3	106	110	107	110
	110_5	110	109	105	110
	110_6	110	105	109	110
possibly yes	103_3	103	105	104	103
	104_8	104	105	109	104
	110_7	110	105	101	110

5 Conclusion and Future Work

We have proposed a new fuzzy identification system that provided the closest three matching candidates for further investigations by experts. The system was built on a center distance weighted local binary pattern. We introduced two new methods: the rotation solver and the core-area to help achieve the best results. The experiments were conducted on the database FVC 2002 DB1-B. The results showed that all proposed new methods in FFIS were successful in detecting whether the fingerprint belonged to the database. In the future, we would like to extend our work with a neural network approach applied to large databases.

References

1. Petkovich, J.: A fingerprint identification system. M.S. thesis, Ottawa-Carleton Institute for Computer Science, Canada, Ottawa, Ontario, 2011. https://doi.org/10.22215/etd/2012-07136. Accessed 26 May 2022

2. Han, Y., Ryu, C., Moon, J., Kim, H., Choi, H.: A study on evaluating the uniqueness of fingerprints using statistical analysis. Lect. Notes Comput. Sci.Comput. Sci. **3506**, 467–477 (2005). https://doi.org/10.1007/11496618_34
3. Turroni, F.: Fingerprint recognition : enhancement, feature extraction and automatic evaluation of algorithms. 24 Ciclo, Ph.D. dissertation, Alma Mater Studiorum Università di Bologna, Bologna (2012). https://doi.org/10.6092/unibo/amsdottorato/4378
4. Sánchez, D., Melin, P., Castillo, O.: Comparison of particle swarm optimization variants with fuzzy dynamic parameter adaptation for modular granular neural networks for human recognition. J. Intell. Fuzzy Syst. **38**(3), 3229–3252 (2020). https://doi.org/10.3233/JIFS-191198
5. Melin, P., Sánchez, D.: Optimal design of type-2 fuzzy systems for diabetes classification based on genetic algorithms. Int. J. Hybrid Intell. Syst. **17**(1–2), 15–32 (2021). https://doi.org/10.3233/his-210004
6. MATLAB—MathWorks—MATLAB & Simulink. https://www.mathworks.com/products/matlab.html. Accessed 21 Jul 2022
7. Chikkerur, S., Wu, C., Govindaraju, V.: LNCS 3072—A Systematic Approach for Feature Extraction in Fingerprint Images. Lecture Notes in Computer Science, Springer, vol. 3072. Berlin, Heidelberg (2007). https://doi.org/10.1007/978-3-540-25948-0_48
8. Tüű-Szabó, B., Kovács, G., Földesi, P., Nagy, S., Kóczy, L.T.: Local Binary Pattern-Based Fingerprint Matching", Computational Intelligence and Mathematics for Tackling Complex Problems 3, pp. 183–188. Springer International Publishing (2022). https://doi.org/10.1007/978-3-030-74970-5_21

Global Sensitivity Analysis and Low Magnitude Pruning for Convolutional Neural Networks Reduction in Image Net Based on Transfer Learning State of the Art Models

Ernest Jeczmionek and Piotr A. Kowalski

1 Introduction

Transfer learning is a method of re-using the stored knowledge of existing models and applying it for related problems. With a growing amount of data and model complexity, a technique is necessary to re-use existing data and solutions [18]. Pruning is a technique for reducing the architecture of a neural network by changing the input or internal parameters of the model. Pruning techniques for neural networks were introduced in the last decade of the twentieth century [6, 10], but due to the current growth of network size, new pruning techniques are being researched and applied [1]. Sensitivity analysis (SA) is a group of methods that connect uncertainty in the model output with respect to the uncertainty of the model input [12]. SA methods have been successfully utilized to minimise the input layer of a neural network such asMLP [3, 19], RBF [13], or PNN [7]. Global Sensitivity Analysis (GSA) is a subgroup of SA. The variance-based GSA used in this article is the Sobol method [11], which modifies all input parameters at the same time so as to identify the most impactful and least impact parameters. Prior to this paper, GSA methods were never used to determine the significance of convolutional filters. This paper advances the GSA method as a tool applicable for pruning convolutional layers.

Low magnitude pruning (LMP) is a well-known method that dates to 80s [9] and which was reintroduced a few years ago [4]. It is a commonly used technique due to

E. Jeczmionek
AGH Doctoral School, AGH University of Krakow, al. A. Mickiewicza 30, 30-059 Cracow, Poland
e-mail: ejec@agh.edu.pl

P. A. Kowalski (✉)
Faculty of Physics and Applied Computer Science, AGH University of Krakow, al. A. Mickiewicza 30, 30-059 Cracow, Poland
e-mail: pkowal@agh.edu.pl

Systems Research Institute, Polish Academy of Sciences, ul. Newelska 6, 01-447 Warsaw, Poland

M. E. Cornejo et al. (eds.), *Computational Intelligence and Mathematics for Tackling Complex Problems 5*, Studies in Computational Intelligence 1127,
https://doi.org/10.1007/978-3-031-46979-4_6

its simplicity. It deletes parameters through a learning period that receives the lowest weight updates [1].

Reduction of neural network structures is a crucial task when the parameter numbers are rapidly growing. One of such areas is forensic analysis, where convolutional networks are commonly applied to classify forensic images [14, 17]. This research reduces the size of some very popular network architectures used in forensic image analysis.

The goal of the research is to investigate the potential of Global Sensitivity Analysis convolution pruning and discern whether a method such as this that generalizes pruning can be competitive with well-established methods like low magnitude pruning. Pre-pruned models can have widespread applications without the need of applying pruning model for a specific application.

2 Experiment

Experiments in this paper were conducted on five state-of-the-art models: VGG16, VGG19, ResNet50V2, ResNet152V2 and IncpetionV3. VGG16 was introduced by Karen Simonyan and Andrew Zisserman in 2014 [15] and employs only 3×3 filters. This network family uses exactly the same architecture topology as AlexNet [8]. VGG16 contains 16 layers and VGG19 has 19 layers. ResNet networks are a residual neural network of a type introduced by Kaiming He for ILSVRC 2015 [5]. Networks such as these are composed of 50 and 152 convolutional layers, with so-called “skip connections” to propagate features to deeper convolutional layers and solve the problem of the vanishing gradients. InceptionV3 architecture, developed by Christian Szegedy for ILSVRC 2014 [16], is a very deep neural network with subnetworks as “inception modules” to reduce the number of internal parameters.

In our experiment, all networks were pre-trained using the ImageNet dataset [2]. Transfer learning was then applied, with frozen convolutional layers and reinitialized parameters within their classification layers, to three most popular datasets: cifar10, cifar100 and fashion MNIST. The goal of this paper is to assess the potential of the GSA method for general convolution pruning and for transfer learning, and to benchmark it against one of the most popular pruning methods—LMP.

For GSA methods, sensitivity of all convolution filters was calculated on a model pre-trained via the ImageNet dataset. The convolution filters were subsequently sorted by applying sensitivity methods, and kernels with the lowest value were set up for pruning. For each final dataset, the classification architecture was not changed, but only randomly initialised. This created a general pruned model, wherein the convolution layers were frozen. In the case of low magnitude pruning, a full transferred model with randomly initialised classification layers was pruned from scratch for each dataset.

3 Results

This section describes the results of comparison low magnitude pruning with GSA on five networks and three datasets. The first network considered is InceptionV3—presented in Fig. 1. It can be noticed that GSA based pruning collapses when 800,000 convolution parameters are removed. Before that, fashion MNIST and cifar10 GSA kept about 80% of their original accuracy. As expected, low magnitude pruning (LMP) preserves test accuracy much better due to the ability to reduce parameters and to prune specific dataset networks.

Figure 2 shows the results for ResNet50V2. It is clear that GSA performs worse than LMP in the case of cifar100, but for light pruning of cifar10, it actually improves original accuracy. More surprising results are observed in the case of fashion MNIST, where a general method like GSA for higher levels of pruning can better sustain the accuracy of the pruning parameters of that method for a specific application. Figure 3 presents the results for the sibling network, that is, ResNet152V2. Similarly to the previous case, LMP sustains stable accuracy, where GSA for the cost of its generality, drops performance at around 5% convolution reduction. Before the sudden drop, the GSA method outperforms LMP for cifar10 and fashion MNIST pushes the test accuracy back to almost its original values. The last decent results for GSA are seen at a level of 96% of origin before the drop. That might suggest that GSA methods can be universally applied up to a certain level of pruning that comes with the generality of different datasets.

The last family of networks is VGG, which are heavy networks with a lot of parameters. The oldest of the two, VGG16, is presented in Fig. 4. While for two

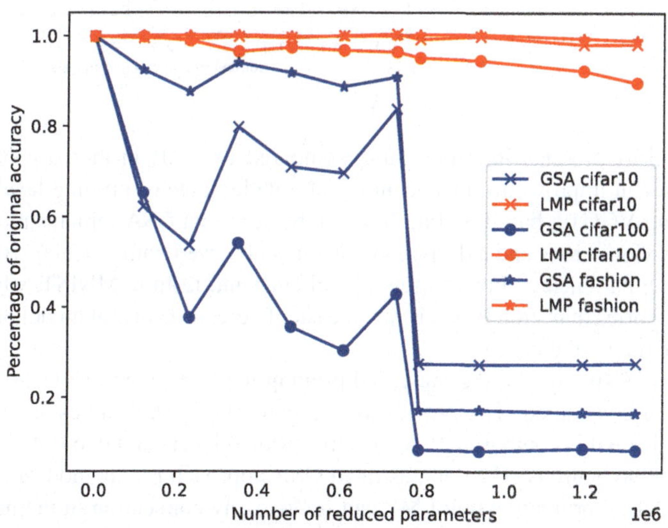

Fig. 1 InceptionV3 10% convolution reduction

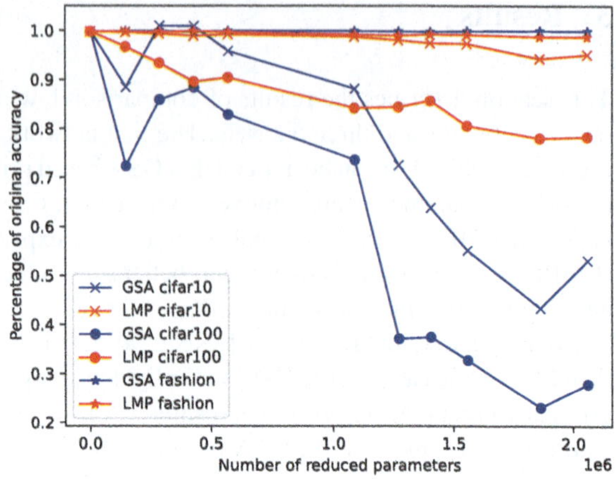

Fig. 2 ResNet50V2 10% convolution reduction

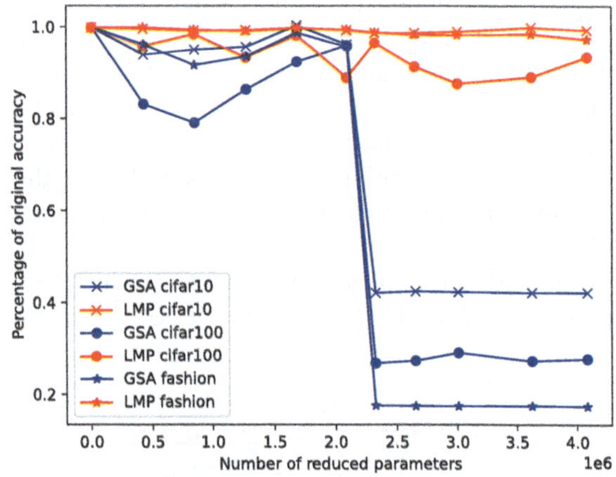

Fig. 3 ResNet152V2 10% convolution reduction

datasets, GSA presents inferior results compared to LMP, in the case of fashion MNIST, it can improve original accuracy at a wide range of pruning levels. When considering VGG19, based on Fig. 5, it can be seen that GSA remains more stable than in previous cases and a drop in precision is observed with reaching 10% of the convolution parameters. For datasets like cifar10 and fashion MNIST, which have fewer categories than cifar100, it is much easier to create a general model of weights for a wide range of datasets.

When choosing one of the presented pruning methods, the model's architecture needs to be considered. The previously presented network families are VGG and ResNet. The first is composed of deep convolutional layers and two fully connected layers of 4096 neurons. The demonstrated GSA approach is a method suitable only for convolution pruning, while LMP can remove any connection in the network—including fully connected layers. This enables LMP to achieve better results in the case of VGG. The ResNet family is more modern than VGG, and is composed mostly

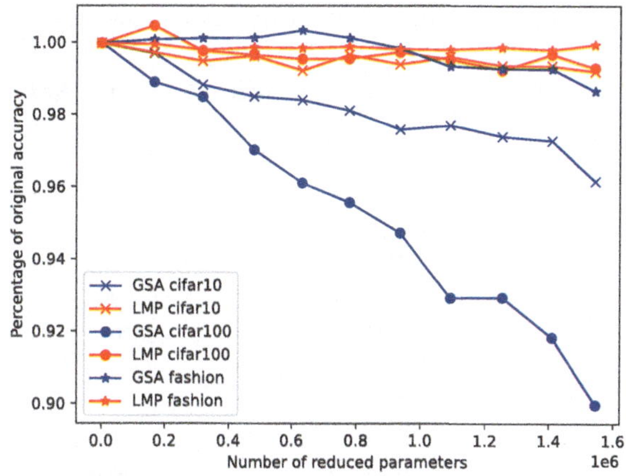

Fig. 4 VGG16 10% convolution reduction

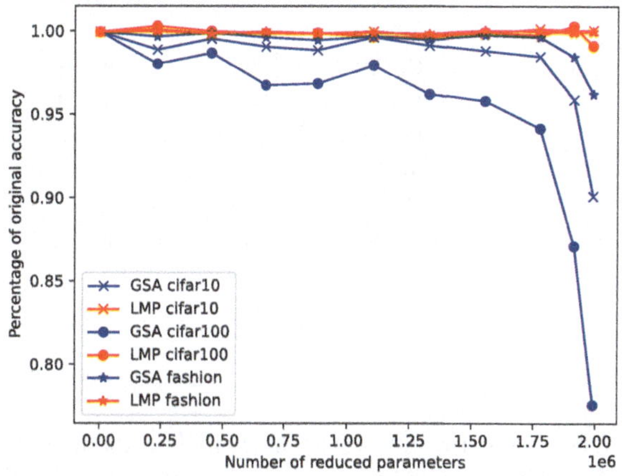

Fig. 5 VGG19 10% convolution reduction

of convolutional layers. Compared to VGG, ResNets provide better accuracy for ImageNet dataset, while being much more memory efficient. ResNet architecture and its derivatives are favourably applied to resolve several current classification problems. The disadvantage of GSA is not an issue if a task does not specifically requires the older architecture.

Table 1 presents best achieved test accuracy for both methods for each network and data set. Results with a red boarder mean that LMP improved the test accuracy of the model, while a blue border indicates that GSA achieved an improvement in accuracy. In five cases, LMP enhanced original accuracy, while GSA did so in 4 examples. LMP increased accuracy in two cases of InceptionV3, while in these cases, GSA did not perform favourably. The situation changes with other networks architectures. GSA enhanced accuracy in half-experiments involving the ResNet family networks, while LMP could only sustain accuracy with a mostly lower level of pruning. In terms of best results, the VGG architecture family is more favourable for LMP in that half of the examples were improved by LSA and GSA made better only one.

Table 1 Best original accuracy percentage and associated reduction for networks and datasets

Network	Dataset	GSA			LMP		
		Best acc. (%)	Param red.	Red. (%)	Best acc. (%)	Param red.	Red. (%)
InceptionV3	CIFAR10	83.9	734490	6	100.4	734490	6
	CIFAR100	65.3	119267	1	99.9	119267	1
	Fashion MNIST	94.1	352654	3	100.3	938265	8
ResNet50V2	CIFAR10	101.1	285999	2	99.9	146224	1
	CIFAR100	88.5	428209	3	96.8	146224	1
	Fashion MNIST	100.1	1861700	9	100.0	428209	3
ResNet152V2	CIFAR10	100.4	1680203	4	100.0%	3616668	9
	CIFAR100	96.0	2090076	5	98.6	845405	2
	Fashion MNIST	98.8	1680203	4	100.0%	423190	1
VGG16	CIFAR10	99.7	165376	1	99.7%	165376	1
	CIFAR100	98.9	165376	1	100.4	165376	1
	Fashion MNIST	100.3	631680	4	99.9%	165376	1
VGG19	CIFAR10	99.6	1105283	5	100.1%	1775122	8
	CIFAR100	98.7	451136	2	100.3%	233088	1
	Fashion MNIST	99.9	451136	2	100.0%	1552143	7

When comparing the best accuracy for both methods, the difference between methods should be remembered. LMP method prunes the whole network specifically for one particular dataset, while GSA creates a general model. That is why the lower number of accuracy improvements was expected for GSA and reaching only one less than dataset-focused pruning is a very good result.

4 Conclusion

Global sensitivity analysis provides a general method for creating a transfer learning model that covers many datasets that involves pruning only the convolutional layers, while the classification layers remain untouched. This convenience comes with the cost of lower accuracy for some datasets, especially that with a larger number of categories. Besides the disadvantage coming from being general, GSA can outperform and even improve the accuracy of a model when compared with methods that prune all types of weights—such as low magnitude pruning. Research shows that the ResNet family is more favourable towards GSA pruning, while LMP performs better with the VGG family—where the older architecture of deeper classification layers is applied. This results in large number of parameters not considered for the presented GSA method. In five cases, LMP improved the accuracy of the model, while GSA did so four times—despite being a general model for broad use. In summary, a single application of GSA pruning on ImageNet pretrained model provides comparable reduction levels to LMP as applied to each case separately. This approach enables generating universal pre-pruned models and reduction of the computational cost of a model pruning in each individual case.

Acknowledgements This article is based upon work from COST Action 17124 DigForAsp, supported by COST (European Cooperation in Science and Technology), www.cost.eu. Research project is partially supported by program Excellence initiative—research university for the AGH University of Krakow, and is partially supported by a Grant for Statutory Activity from the Faculty of Physics and Applied Computer Science of the AGH University of Krakow. The authors would also like to thank the anonymous referees for their careful reading of the paper and their contribution of useful suggestions that helped to improve this article.

References

1. Blalock, D., Ortiz, J.J.G., Frankle, J., Guttag, J.: What is the State of Neural Network Pruning? (2020)
2. Deng, J., Dong, W., Socher, R., Li, L.J., Li, K., Fei-Fei, L.: Imagenet: A large-scale hierarchical image database. In: 2009 IEEE Conference on Computer Vision and Pattern Recognition, pp. 248–255. IEEE (2009)
3. Fock, E.: Global sensitivity analysis approach for input selection and system identification purposes-a new framework for feedforward neural networks. IEEE Trans. Neural Netw. Learn. Syst. **25**, 1484–1495 (2014). https://doi.org/10.1109/TNNLS.2013.2294437
4. Han, S., Pool, J., Tran, J., Dally, W.J.: Learning both weights and connections for efficient neural networks. In: Proceedings of the 28th International Conference on Neural Information Processing Systems, vol. 1, pp. 1135–1143. NIPS'15, MIT Press, Cambridge, MA, USA (2015)
5. He, K., Zhang, X., Ren, S., Sun, J.: Deep residual learning for image recognition. In: Proceedings of the IEEE Conference on Computer Vision and Pattern Recognition, pp. 770–778 (2016)
6. Karnin, E.: A simple procedure for pruning back-propagation trained neural networks. IEEE Trans. Neural Netw. **1**(2), 239–242 (1990). https://doi.org/10.1109/72.80236
7. Kowalski, P.A., Kusy, M.: Determining significance of input neurons for probabilistic neural network by sensitivity analysis procedure. Comput. Intell. **34**(3), 895–916 (2018)
8. Krizhevsky, A., Sutskever, I., Hinton, G.E.: Imagenet classification with deep convolutional neural networks. Adv. Neural Inf. Process. Syst. **25**, 1097–1105 (2012)
9. Lecun, Y., Denker, J., Solla, S.: Optimal Brain Damage, vol. 2, pp. 598–605 (1989)
10. Mozer, M.C., Smolensky, P.: Skeletonization: A technique for trimming the fat from a network via relevance assessment. In: Proceedings of the 1st International Conference on Neural Information Processing Systems, pp. 107–115. NIPS'88, MIT Press, Cambridge, MA, USA (1988)
11. Saltelli, A., Annoni, P., Azzini, I., Campolongo, F., Ratto, M., Tarantola, S.: Variance based sensitivity analysis of model output design and estimator for the total sensitivity index. Comput. Phys. Commun. **181**, 259–270 (2010). https://doi.org/10.1016/j.cpc.2009.09.018
12. Saltelli, A., Tarantola, S., Campolongo, F., Ratto, M.: Sensitivity Analysis in Practice: A Guide to Assessing Scientific Models (2004). https://doi.org/10.1002/0470870958

13. Shi, D., Yeung, D.S., Gao, J.: Sensitivity analysis applied to the construction of radial basis function networks. Neural Netw. **18**(7), 951–957 (2005)
14. Silva, S.H., Bethany, M., Votto, A.M., Scarff, I.H., Beebe, N., Najafirad, P.: Deepfake forensics analysis: An explainable hierarchical ensemble of weakly supervised models. Forensic Sci. Int.: Synergy **4**, 100217 (2022)
15. Simonyan, K., Zisserman, A.: Very deep convolutional networks for large-scale image recognition (2014). arXiv:1409.1556
16. Szegedy, C., Liu, W., Jia, Y., Sermanet, P., Reed, S., Anguelov, D., Erhan, D., Vanhoucke, V., Rabinovich, A.: Going deeper with convolutions. In: 2015 IEEE Conference on Computer Vision and Pattern Recognition (CVPR), pp. 1–9 (2015). https://doi.org/10.1109/CVPR.2015.7298594
17. Yang, P., Baracchi, D., Ni, R., Zhao, Y., Argenti, F., Piva, A.: A survey of deep learning-based source image forensics. J. Imaging **6**(3) (2020). https://doi.org/10.3390/jimaging6030009, https://www.mdpi.com/2313-433X/6/3/9
18. Yosinski, J., Clune, J., Bengio, Y., Lipson, H.: How transferable are features in deep neural networks? Adv. Neural Inf. Process. Syst. (NIPS) **27** (2014)
19. Zurada, J.M., Malinowski, A., Usui, S.: Perturbation method for deleting redundant inputs of perceptron networks. Neurocomputing **14**(2), 177–193 (1997)

Generalized Linearity of Aggregation and Related Functions

Radko Mesiar and Andrea Stupňanová

1 Introduction

A linearization is a substantial tool for creating an appropriate and admissible mathematical model for many practical problems. The best local linear approximation of a 1-dimensional function f in point x_0 is its tangent line t given by $t(x) = f(x_0) + f'(x_0)(x - x_0)$ if the derivative exists. Similarly, for 2-dimensional functions we consider tangent planes, and if $n > 2$, for n-dimensional functions the related tangent hyperplanes are considered. In the case, when a global approximation has to be proposed, linear splines are standardly considered, and their continuity is a classical constraint.

The aim of this contribution is the study of particular linear splines defined on $[0, 1]^n$ with outputs from $[0, 1]$, i.e., of continuous piecewise linear fusion function, see e.g. [4]. Consider a finite simplex division $\mathbb{S} = \{S_1, \ldots, S_k\}$ of $[0, 1]^n$ (i.e., $\lambda_n(S_i \cap S_j) = 0, i \neq j$ for any $i, j = 1, \ldots, k$, where λ_n is the standard Lebesgue measure on Boolean subsets of $[0, 1]^n$, (i.e., ordinary crisp subsets), and $\bigcup_{i=1}^k S_i = [0, 1]^n$). The considered piecewise linear function F is supposed to be linear on each simplex S_i, i.e.,

$$F(x_1, \ldots, x_n) = \sum_{j=1}^{n} a_{i,j} x_j + b_i \ \in [0, 1]$$

This work was supported by the Slovak Research and Development Agency under the contract no. APVV-17-0066 and no. APVV-18-0052 and grant VEGA 1/0006/19.

R. Mesiar · A. Stupňanová (✉)
Faculty of Civil Engineering, Slovak University of Technology in Bratislava, Bratislava, Slovak Republic
e-mail: andrea.stupnanova@stuba.sk

R. Mesiar
e-mail: radko.mesiar@stuba.sk

M. E. Cornejo et al. (eds.), *Computational Intelligence and Mathematics for Tackling Complex Problems 5*, Studies in Computational Intelligence 1127,
https://doi.org/10.1007/978-3-031-46979-4_7

for each $(x_1, \ldots, x_n) \in S_i, i = 1, \ldots, k$.

In the next section, we discuss some general properties and some particular constructions for piecewise linear continuous fusion functions. In Sect. 3, we focus on dimension $n = 2$ and discuss situations dealing with $k = 2$ of considered simplices. In Sect. 4, a particular example based on 4-simplices is presented. Finally, some concluding remarks are added.

2 Continuous Piecewise Linear Fusion and Aggregation Functions

Fix an integer n. Recall that a fusion function is any mapping $F : [0, 1]^n \to [0, 1]$ (i.e., $([0, 1], F)$ can be see as an n-ary grupoid). For more details and applications we recommend [4, 10]. The set of all fusion functions we denote as $\mathcal{F}$ (i.e., $F \in \mathcal{F}$). If a fusion function F satisfies the boundary conditions $F(\mathbf{0}) = 0$ and $F(\mathbf{1}) = 1$, then F is called a *semi-aggregation function*, $(F \in \mathcal{S})$.

If a semi-aggregation function F

- satisfies $F(\mathbf{x} + \mathbf{c}) \geq F(\mathbf{x})$ for any $\mathbf{x}, \mathbf{x} + \mathbf{c} \in [0, 1]^n$, $\mathbf{c} = (c, \ldots, c)$, $c > 0$, (i.e., F is weakly increasing), then F is called a *weak aggregation function*, $(F \in \mathcal{W})$, see [13];
- for a fixed non-zero vector $\overrightarrow{\mathbf{r}} = (r_1, \ldots, r_n) \in [0, 1]^n$ satisfies $F(\mathbf{x} + c\overrightarrow{\mathbf{r}}) \geq F(\mathbf{x})$ for any $\mathbf{x}, \mathbf{x} + c\overrightarrow{\mathbf{r}} \in [0, 1]^n$, $c > 0$, (i.e., F is $\overrightarrow{\mathbf{r}}$-directionally increasing), then F is called a *pre-aggregation function*, $(F \in \mathcal{P})$, see [10];
- is monotone increasing in each coordinate, (i.e., F is $\vec{e}_i$-increasing for each $i = 1, \ldots, n$, $\vec{e}_i = (0, \ldots, \underbrace{1}_{i\text{-th coordinate}}, \ldots, 0)$), then F is called an *aggregation function*, $(F \in \mathcal{A})$, see [3, 8].

Note that $\mathcal{P}$ is the class of all pre-aggregation functions, independently of vector $\overrightarrow{\mathbf{r}}$. For fixed vector $\overrightarrow{\mathbf{r}}$ the class of all $\overrightarrow{\mathbf{r}}$-directionally increasing semi-aggregation function we denote by $\mathcal{S}_{\overrightarrow{\mathbf{r}}}$, and finally we denote by $\mathcal{F}_*$ the class of all $\overrightarrow{\mathbf{r}}$-directionally increasing continuous fusion functions for some vector $\overrightarrow{\mathbf{r}}$.

Continuous piecewise linear functions from these classes form the subclasses $\mathcal{LF}, \mathcal{LS}, \mathcal{LW}, \mathcal{LF}_*, \mathcal{LS}_{\overrightarrow{\mathbf{r}}}, \mathcal{LP}$ and $\mathcal{LA}$. Both systems can be ordered by subsethood relation, and then their Hasse diagrams, see Fig. 1, coincide.

Theorem 1 *Let m, n be fixed integers. Then for any m-dimensional fusion function F and $G_1, \ldots, G_m$ n-dimensional fusion functions also the composite $H : [0, 1]^n \to [0, 1]$ given by $H(x_1, \ldots, x_n) = F(G_1(x_1, \ldots, x_n), \ldots, G_m(x_1, \ldots, x_n))$ is a fusion function. Similar claim holds for semi-aggregation and aggregation functions, but not for the other introduced classes, in general.*

Observe that the continuity and piece-wise linearity is preserved by composition of functions, thus yielding the next result.

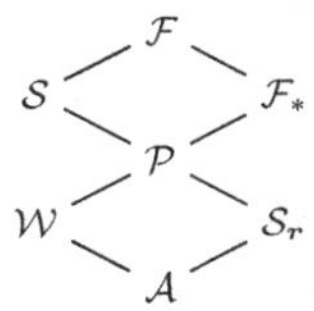

Fig. 1 Hasse diagram for classes $\mathcal{F}, \mathcal{S}, \mathcal{W} \ldots$, where $\overrightarrow{\mathbf{r}} \neq (c, \ldots, c)$

Corollary 1 *Let m, n be a fixed integers. Under the notation of Theorem 1, and considering all functions $F, G_1, \ldots, G_m$ to be continuous piece-wise linear fusion (semi-aggregation, aggregation) functions, also the composite H is a continuous piece-wise linear fusion (semi-aggregation, aggregation) function.*

As a particular case, one can consider either $n = 1$ or $m = 1$. When combining consecutively both cases and considering 1-dimensional fusion function $\varphi : [0, 1] \to [0, 1], \varphi(x) = 1 - x$, for any fusion function $F : [0, 1]^n \to [0, 1]$ its dual $F^d : [0, 1]^n \to [0, 1]$ given by $F^d(x_1, \ldots, x_n) = 1 - F(1 - x_1, \ldots, 1 - x_n)$ is obtained and it is also a fusion function. Clearly,

$$F \in \mathcal{S} \text{ if and only if } F^d \in \mathcal{S}; \qquad F \in \mathcal{W} \text{ if and only if } F^d \in \mathcal{W};$$
$$F \in \mathcal{P} \text{ if and only if } F^d \in \mathcal{P}; \qquad F \in \mathcal{A} \text{ if and only if } F^d \in \mathcal{A}.$$

Typical piece-wise linear aggregation functions are the well known Choquet [5], Sugeno [11] and Shilkret [12] integrals, but also normed decomposition integrals proposed by Even and Lehrer [7]. Recall that, considering a probability measure on $\{1, \ldots, n\}$ characterized by the probability vector $(p_1, \ldots, p_n)$, the related weighted arithmetic mean is given by $\text{WAM}_{\mathbf{p}}(\mathbf{x}) = \sum_{i=1}^{n} p_i x_i$ (i.e., it is an expected value of the random variable $\mathbf{x}$). Based on Corollary 1, one can construct new types of linear splines related to aggregation functions considering 2-step Choquet integrals (k-step Choquet integrals, $k > 2$) or multistep integrals introduced by Narukawa and Torra [9]. Also, some discussion concerning monotone increasing linear splines due to Torra and Narukawa [9] is of interest.

3 2-Simplices Based Case

Fix $n = 2$, i.e., we will consider continuous piece-wise linear fusion function $F : [0, 1]^2 \to [0, 1]$. The simplest example is $F \equiv c, c \in [0, 1]$, which is a monotone linear fusion, but not a semi-aggregation function. Observe that 2 is the smallest number of simplices forming a division of $[0, 1]^2$, and in such a case there are two possible situations:

(i) $S_1^+ = \triangle\ ((0, 0), (1, 0), (1, 1))$ is the triangle determined by indicated 3 vertices, and $S_2^+ = \triangle\ ((0, 0), (0, 1), (1, 1))$, see Fig. 2 left;

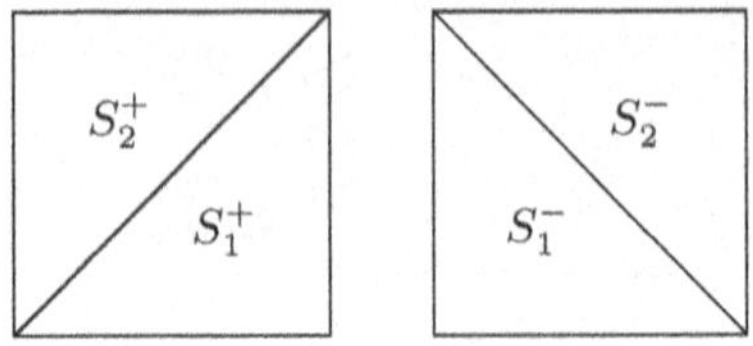

Fig. 2 Two possible simplex divisions of $[0, 1]^2$ into 2 simplices

(ii) $S_1^- = \triangle\,((0, 0), (1, 0), (0, 1))$ and $S_2^- = \triangle\,((1, 0), (1, 1), (0, 1))$, see Fig. 2 right.

To determine a linear function on some simplex S, we need (and it is enough) to know the values of the considered function on vertices of S. Then, in the discussed case, we need to know the values $F(0, 0) = a$, $F(1, 0) = b$, $F(0, 1) = c$ and $F(1, 1) = d$. Then, for $a, b, c, d \in [0, 1]$,

(i) if $\mathbb{S} = (S_1^+, S_2^+)$ is considered, then

$$F(x, y) = \begin{cases} x(b-a) + y(d-b) + a & \text{if } (x, y) \in S_1^+, \\ x(d-c) + y(c-a) + a & \text{if } (x, y) \in S_2^+. \end{cases}$$

The function F is a fusion function which is continuous and piece-wise linear (i.e., a linear spline on $[0, 1]^2$). For special constants a, b, c, d, it holds:

- $F \in \mathcal{S}$ if and only if $a = 0$ and $d = 1$, and then F is the Choquet integral with respect to the fuzzy measure $\mu : 2^{\{1,2\}} \to [0, 1]$, $\mu(\{1\}) = b$ and $\mu(\{2\}) = c$. Hence, F is an aggregation function. More, if $b = c$, then F is an OWA operator [14]. If $b = c = 0$ then $F = \min$, if $b = c = 1$ then $F = \max$ and if $b = c = \frac{1}{2}$ then $F = \text{AM}$. F is supermodular (submodular) if $b + c \geq 1$ $(b + c \leq 1)$;
- F is weakly increasing if and only if $d \geq a$;
- F is monotone increasing if and only if $a \leq b \leq d, a \leq c \leq d$;
- if $a = c = d = 1$ and $b = 0$ then F is the Lukasiewicz implication [2]. Similarly, all other binary Boolean operations can be extended into basic linear splines based on simplices (S_1^+, S_2^+). Then conjuction (AND) is extended into $T_M = \min$ on $[0, 1]^2$, disjunction (OR) into $S_M = \max$ on $[0, 1]^2$, Sheffer stroke (NAND) into $1 - T_M$, etc., see [1].

(ii) if $\mathbb{S} = (S_1^-, S_2^-)$ is considered, then

$$F(x, y) = \begin{cases} x(b-a) + y(c-a) + a & \text{if } (x, y) \in S_1^-, \\ x(d-c) + y(d-b) + b + c - d & \text{if } (x, y) \in S_2^-. \end{cases}$$

The function F is a continuous piece-wise linear (i.e., a linear spline) fusion function. For special constants a, b, c, d, it holds:

- $F \in \mathcal{S}$ if and only if $a = 0$ and $d = 1$, and then F is an aggregation function; it is a weighted arithmetic mean if $b + c = 1$ and if $b + c \leq 1$ ($b + c \geq 1$) then F is supermodular (submodular);
- F is weakly increasing if and only if $a \leq \frac{b+c}{2} \leq d$;
- F is monotone increasing if and only if $a \leq b \leq d$, and $a \leq c \leq d$;
- if $a = c = d = 1$ and $b = 0$ then F is the Kleene-Dienes implication [2];
- considering the extensions of the other Boolean operations, AND is extended into the Lukasiewitz t-norm T_L, OR into the Lukasiewitz t-conorm S_L, NAND into $1 - T_L$, etc.

Note that for $n \geq 2$ and $n!$-simplices generated by permutation on $(1, \dots, n)$, supposing the monotonicity of the related F defined on vertices from $\{0, 1\}^n$ and $F(\mathbf{0}) = 0$, $F(\mathbf{1}) = 1$, F on $[0, 1]^n$ is a linear spline if and only if it is the Choquet integral with respect to the fuzzy measure $\mu : 2^{\{1,\dots,n\}} \to [0, 1]$ given by $\mu(E) = F(1_E)$.

4 4-Simplices Based Example

Note that for any $n \in \mathbb{N}$, the smallest number k of non-overlapping simplices is a simplex division of $[0, 1]^n$ is $n!$. Thus, if $k \in \{2, 3, 4, 5\}$, necessarily $n = 2$. We focus now on a particular case $k = 4$ (and thus $n = 2$.)

Example 1 As an interesting example considering 4 simplices S_1, S_2, S_3 and S_4, see Fig. 3, determined by a point $(u, v) \in]0, 1[^2$ and 4 vertices of $[0, 1]^2$ suppose $F(u, v) = a$, $F(0, 0) = F(1, 0) = F(0, 1) = 0$ and $F(1, 1) = 1$, i.e., $F|\{0, 1\}^2$ is the Boolean conjunction. Then

$$F(x, y) = \begin{cases} \frac{a}{v} y & \text{if } (x, y) \in S_1, \\ \frac{a}{u} x & \text{if } (x, y) \in S_2, \\ \frac{u-a}{1-v}(y - 1) + x & \text{if } (x, y) \in S_3, \\ \frac{v-a}{1-u}(x - 1) + y & \text{if } (x, y) \in S_4. \end{cases} \tag{1}$$

The function F given by (1) is a fusion function for any possible $(u, v) \in]0, 1[^2$ and $a \in [0, 1]$. Also, always $F \in \mathcal{S}$. More:

- $F \in \mathcal{A}$ if and only if $a \leq \min(u, v)$, and then F is a semicopula [6];
- F is a quasi-copula [6] if and only if F is a copula [6] if and only if $\max(0, u + v - 1) \leq a \leq \min(u, v)$ (and then F is supermodular).

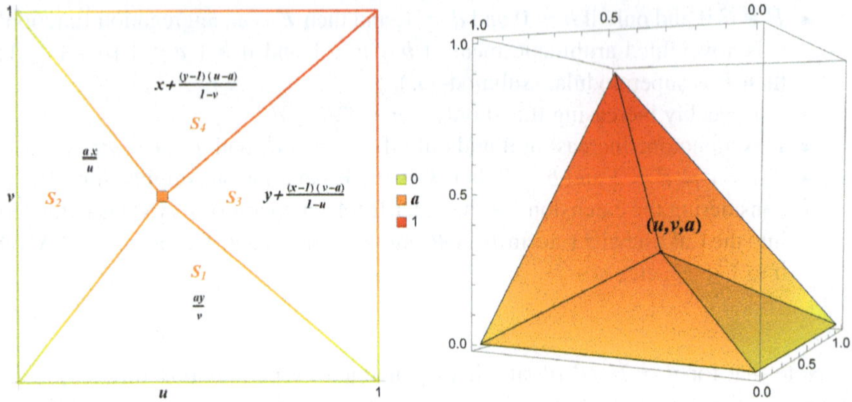

Fig. 3 Description of Example 1

5 Concluding Remarks

After recalling of fusion functions and several special subclasses of them, we have studied piece-wise linear continuous fusion functions. These can be seen as linear splines related to a finite number k of non-overlapping simplices covering the full domain $[0, 1]^n$. To determine such linear splines we need to know only a final number of studied model values attained or estimated in vertices of those k-simplices. We have introduced some general properties and some construction methods and completely characterized the case when $n = 2$ and the minimal number $k = 2$ of simplices is considered. Also, an interesting example dealing with 4-simplices (and thus necessarily $n = 2$) was also presented. For the further study of generalized linearity observe that positively homogeneous n-ary fusion function is linear on any segment connecting points $\mathbf{0}$ and $\mathbf{x}$ from $[0, 1]^n$. This idea to require linearity (affinity) on some subsets of $[0, 1]^n$ (closed, convex) which are non-overlapping and covering $[0, 1]^n$ and which need not be simplices is an interesting topic for the further research.

References

1. Baczynski, M., et al.: On the Sheffer stroke operation in fuzzy logic. Fuzzy Sets Syst. **431**, 110–128 (2022)
2. Baczynski, M., Jayaram, B.: Fuzzy Implications. Studies in Fuzziness and Soft Computing, vol. 231. Springer, Berlin (2008)
3. Beliakov, G., Pradera, A., Calvo, T.: Aggregation Functions: A Guide for Practitioners. Springer, Berlin (2007)
4. Bustince, H., Fernández, J., Kolesárová, A., Mesiar, R.: Directional monotonicity of fusion functions. Eur. J. Oper. Res. **244**(1), 300–308 (2015)
5. Choquet, G.: Theory of capacities. Ann. Inst. Fourier **5**, 131–295 (1953/1954)
6. Durante, F., Sempi, C.: Principles of Copula Theory. CRC Press, Florida (2015)

7. Even, Y., Lehrer, E.: Decomposition-integral: unifying choquet and the concave integrals. Econ. Theory **56**(1), 1–26 (2014)
8. Grabisch, G., Marichal, J.-L., Mesiar, R., Pap, E.: Aggregation Functions. Cambridge University Press, Cambridge (2009)
9. Narukawa, Y., Torra, V.: Twofold integral and multi-step Choquet integral. Kybernetika **40**(1), 39–50 (2004)
10. Lucca, G., Sanz, J.A., Dimuro, G.P., Bedregal, B., Mesiar, R., Kolesárová, A., Bustince, H.: Pre-aggregation functions: construction and an application. IEEE Trans. Fuzzy Syst. **24**(2), 260–272 (2016)
11. Sugeno, M.: Theory of fuzzy integrals and its applications. Ph.D. thesis, Tokyo Institute of Technology (1974)
12. Shilkret, N.: Maxitive measure and integration. Indag. Math. **33**, 109–116 (1971)
13. Wilkin, T., Beliakov, G.: Weakly monotone averaging functions. Int. J. Intell. Syst. **30**, 144–169 (2015)
14. Yager, R.R.: On ordered weighted averaging aggregation operators in multicriteria decision-making. IEEE Trans. Syst. Man Cybern. **18**(1), 183–190 (1988)

10. [illegible]
11. [illegible]
12. [illegible]
13. [illegible]
14. [illegible]

Closure Operators on AMV-Valued Fuzzy Sets

Jiří Močkoř

1 Introduction

In contemporary fuzzy mathematics and its applications, there are several general trends that characterize the directions of their development. These trends undoubtedly include the development of new fuzzy structures with values in various partially ordered algebraic structures. Let us mention, for example, intuitionistic fuzzy sets [2, 3], fuzzy soft sets [6, 11–14], or neutrosophic fuzzy sets [9, 10] and their mutual combinations, such as intuitionistic fuzzy soft sets [1, 7], hesitant fuzzy soft sets [4, 19, 20], and many others. Similarly, with this trend of new fuzzy structures, opposite efforts are being made to unify the methods used by these new fuzzy structures. One of the possible obstacles to unifying the theories of these new L-fuzzy structures, where L is a complete MV-algebra, is that although these new fuzzy structures are also traditionally called L-fuzzy sets, they are not actually mappings $X \to L$. A typical example is the theory of L-fuzzy soft sets in space (X, K), where X is the basic set and K is the fixed set of criteria. In that case, a L fuzzy soft set is a pair (E, s), where $E \subseteq K$ and $s : E \to L^X$. Subsequently, this complicates the effort to apply the methods commonly used in classic L-fuzzy sets for these new L-fuzzy structures.

In our previous papers [15, 16] we showed that many of these new L-fuzzy structures, where L is a complete MV-algebra, can be transformed to $\mathcal{F}(L)$-fuzzy sets $X \to \mathcal{F}(L)$, where $\mathcal{F}(L)$ is an algebraic structure called the complete almost MV-algebra (complete AMV-algebra, shortly). This transformation of the new

This work was partly supported from ERDF/ESF project CZ.02.1.01/0.0/0.0/17-049/0008414.

J. Močkoř (✉)
Institute for Research and Applications of Fuzzy Modeling, University of Ostrava,
Centre of Excellence IT4Innovations, 30. dubna 22, 701 03 Ostrava, Czech Republic
e-mail: Jiri.Mockor@osu.cz
URL: http://irafm.osu.cz/

M. E. Cornejo et al. (eds.), *Computational Intelligence and Mathematics for Tackling Complex Problems 5*, Studies in Computational Intelligence 1127,
https://doi.org/10.1007/978-3-031-46979-4_8

L-fuzzy structures into a uniform type of $\mathcal{F}(L)$-fuzzy sets subsequently enables the use of classic fuzzy set methods for these new L-fuzzy structures as well, including various operations defined on the powerset objects of these new fuzzy structures.

The second development trend of contemporary fuzzy set theory is the use of algebraic and topological methods applied to the powerset objects of individual fuzzy structures. The basic idea of this approach was the definition of a new theory that contains among powerset objects also the topology defined on these powerset objects and the inclusion of relevant axioms concerning continuity of maps to this new theory (see, e.g., Rodabaugh [17], Solovyov [18] with an extended list of references). This new theory was called *the topological powerset theory*. This approach made it possible to apply topological methods to other powerset structures in comparison to the traditional powerset objects of crisp or fuzzy sets and also to better understand the algebraic character of these structures, using tools from the theory of monads.

In this paper, we want to combine both trends and to show how closure structures can be defined for new types of L-fuzzy structure using unification based on the application of transformation of these new L-fuzzy structures to AMV-valued fuzzy sets.

2 AMV-Valued Fuzzy Sets

As we mentioned in the Introduction, basic tools we use in our paper are AMV-algebras. In the paper [16] we prove that the category of complete AMV-algebras is isomorphic to the category of some pairs of idempotent commutative semirings. In this section we repeat basic definitions related to these notions, we introduce some basic properties of AMV-valued fuzzy sets and the definition of partially ordered monads that we will use for our constructions.

Definition 1 ([16]) The complete AMV-algebra is the algebra $(R, +, \times, \neg, 0, 1)$ of type $(2, 2, 1, 0, 0)$ that satisfies the following axioms:

1. $(R, +, 0)$ is the complete idempotent commutative monoid, the sum of elements from $S \subseteq R$ is indicated by $\sum_{r \in S} r$,
2. $(R, \times, 1)$ is a commutative monoid,
3. $\neg : R \to R$ is the involution mapping, $\neg 0 = 1$,
4. $(\forall x, y_i \in R, i \in I), \quad x \times \sum_i y_i = \sum_i (x \times y_i)$,
5. $(\forall x \in R), \quad x \times 0 = 0$,
6. For $x, y, z, y_i \in R, i \in I$,

   (a) $(x +^* y = x) \Leftrightarrow (x + y = y)$,
   (b) $x \times^* \sum_{i \in I} y_i = \sum_{i \in I} x \times^* y_i$,
   (c) $x + (y +^* z) = (x + y) +^* (x + z)$,

   where for $x, y \in R$, $x +^* y := \neg(\neg x + \neg y), \quad x \times^* y := \neg(\neg x \times \neg y)$.

The properties of the AMV-algebra are described in the following lemma.

Lemma 1 ([16]) *Let $\mathcal{R} = (R, +, \times, \neg, 0, 1)$ be the complete AMV-algebra.*

1. *Let the relations $\leq$ and $\leq^*$ be defined by*

$$x, y \in R, \quad x \leq y \Leftrightarrow x + y = y, \quad x \leq^* y \Leftrightarrow x +^* y = y.$$

The following statements hold:

(a) *$\leq$ and $\leq^*$ are the order relations on R,*
(b) $x \leq^* y \Leftrightarrow x \geq y$,
(c) $x \geq y \Leftrightarrow \neg x \leq \neg y$,
(d) $x \leq y \Rightarrow x + z \leq y + z, \quad x +^* z \leq z +^* z$,
(e) $x \leq y \Rightarrow x \times z \leq y \times z, \quad x \times^* y \leq y \times^* z$.

2. *$(R, \leq)$ and $(R, \leq^*)$ are complete lattices, where, for arbitrary $S \subseteq R$,*

$$\sup S = \sum_{x \in S} x, \quad \inf S = \sum_{x \in S}^{*} x, \; in\ (R, \leq),$$

$$\sup S = \sum_{x \in S}^{*} x, \quad \inf S = \sum_{x \in S}, \; in\ (R, \leq^*).$$

where $\sum_{x \in S}^{*} x$ is the sum of elements with respect to $+^*$.

3. *Any complete MV-algebra $(L, \oplus, \neg, 0)$ is a complete AMV-algebra, but the converse does not hold.*

Using AMV-algebras, we can also introduce AMV-valued fuzzy sets and operations with these fuzzy sets.

Definition 2 ([16]) Let $\mathcal{R} = (R, +, \times, \neg, 0, 1)$ be the complete AMV algebra, and let X be a set.

1. A mapping $s : X \to R$ is called an AMV-valued fuzzy set in a set X with values in $\mathcal{R}$ ($\mathcal{R}$-fuzzy set, shortly).
2. Operations with $\mathcal{R}$-fuzzy sets are defined by

   (a) The intersection $s \sqcap t$ is defined by $(s \sqcap t)(x) = s(x) +^* t(x), x \in X$,
   (b) The union $s \sqcup t$ is defined by $(s \sqcup t)(x) = s(x) + t(x), x \in X$,
   (c) Complement $\neg s$ is defined by $(\neg s)(x) = \neg(s(x))$,
   (d) The external multiplication $\star$ by elements of R is defined by $(a \star s)(x) = a \times s(x)$,
   (e) The order relation $\leq$ between s, t is defined by $s \leq t \Leftrightarrow (\forall x \in X) s(x) \leq t(x)$.

As we mentioned in the Introduction, the AMV-algebra can be equivalently defined using a pair of commutative idempotent semirings, among which there is an involutive isomorphism. This equivalent definition of the AMV-algebra becomes especially useful in situations where pairs of operations are used, such as upper and lower approximations of fuzzy sets using fuzzy relations or lower and upper F-transforms. In the paper, we use this equivalent form of AMV-algebras.

The following definition of dual pairs of semirings was introduced in [16].

Definition 3 Let $\mathcal{R} = (R, +, \times, 0, 1)$ and $\mathcal{R}^* = (R, +^*, \times^*, 0^*, 1^*)$ be complete idempotent commutative semirings with the same underlying set R. The pair $(\mathcal{R}, \mathcal{R}^*)$ is called the dual pair of semirings if there exists a mapping $\neg : R \to R$ and the following axioms hold:

1. $\neg : \mathcal{R} \to \mathcal{R}^*$ is the involutive isomorphism of the semirings,
2. $\forall a \in R, S \subseteq R \quad a \times^* (\sum_{b\in S} b) = \sum_{b\in S}(a \times^* b)$,
3. $\forall a \in R, S \subseteq R \quad a + (\sum^*_{b\in S} b) = \sum^*_{b\in S}(a + b)$, where $\sum^*$ is the complete operation $+^*$ in $\mathcal{R}^*$,
4. $\forall a, b \in R, \quad a + b = a \Leftrightarrow a +^* b = b$.

Using the dual isomorphism $\neg$, it is easy to see that the following dual statements also hold for arbitrary dual pairs of semirings $(\mathcal{R}, \mathcal{R}^*)$:

2'. $\forall a, \in R, S \subseteq R, \quad a \times \sum^*_{b\in S} b = \sum^*_{b\in S}(a \times b)$,
3'. $\forall a, \in R, S \subseteq R, \quad a +^* \sum_{b\in S} b = \sum_{b\in S}(a +^* b)$.

In [16] we introduce the category of AMV-algebras **AMVAlg** and the category of dual pairs adjoint semirings **DPSem** and prove that these categories are isomorphic.

Definition 4 1. The category **DPSem** of dual pairs of semirings is defined by

(a) Objects $((\mathcal{R}, \mathcal{R}^*), \neg_{\mathcal{R}})$ are dual pairs $(\mathcal{R}, \mathcal{R}^*)$ of semirings with involutive isomorphism $\neg_{\mathcal{R}} : \mathcal{R} \to \mathcal{R}^*$,
(b) Morphisms $((\mathcal{R}, \mathcal{R}^*), \neg_{\mathcal{R}}) \to ((\mathcal{S}, \mathcal{S}^*), \neg_{\mathcal{S}})$ are mappings between underlying sets $\varphi : R \to S$, such that
   i $\varphi : \mathcal{R} \to \mathcal{S}$ and $\varphi : \mathcal{R}^* \to \mathcal{S}^*$ are homomorphisms of semirings,
   ii The identity $\neg_{\mathcal{S}}.\varphi = \varphi.\neg_{\mathcal{R}}$ holds.
(c) The compositions of the morphisms are the standard compositions of the mappings.

2. The category **AMVAlg** of complete AMV-algebras is defined by

(a) Objects are complete AMV-algebras $\mathcal{R} = (R, +, \times, \neg, 0, 1)$,
(b) Morphisms $\psi : (R, +, \times, \neg, 0, 1) \to (S, +_1, \times_1, \neg_1, 0_1, 1_1)$ are mappings $\psi : R \to S$ such that ψ is a homomorphism of these algebras.
(c) The composition of morphisms is the standard composition of mappings.

Remark 1 From the existence of the isomorphism F of these categories, it follows that if L is a AMV-algebra, the L-fuzzy set can be equivalently called the $(\mathcal{R}, \mathcal{R}^*)$-fuzzy set, where $F(L) = ((\mathcal{R}, \mathcal{R}^*), \neg)$. We use this terminology of $(\mathcal{R}, \mathcal{R}^*)$-fuzzy sets in the remainder of this paper.

For illustration, we present two examples of transformations of new L-fuzzy structures into $\mathcal{F}(L)$-fuzzy set, where $L = (L, \vee, \wedge, \oplus, \otimes, \neg, 0_L, 1_L)$ is a complete MV-algebra.

Example 1 Recall [11] that if X is a set of objects to be evaluated and K is a set of all possible criteria, then a pair (E, s) is called an L-fuzzy soft set in space (X, K), if $E \subseteq K$ and $s : E \to L^X$. We show how the L-fuzzy soft set can be transformed into the $(\mathcal{R}, \mathcal{R}^*)$ -fuzzy set. The structure $(\mathcal{R}, \mathcal{R}^*)$ is defined by

1. The semiring $\mathcal{R} = (R, +, \times, 0, 1)$ is defined by
   (a) $R = \{(E, s) : E \subseteq K, s \in L^K\} \subseteq L^K$, where $(E, s) \in L^K$ is defined by

$$k \in K, \quad (E, s)(k) = \begin{cases} s(k), & k \in E, \\ 0_L, & k \notin E \end{cases}.$$

   (b) $(E, t), (F, s) \in R$, $(E, t) + (F, s) := (E \cap F, t \vee s)$, where $t \vee s$ is the supremum in L^K,
   (c) $(E, t), (F, s) \in R$, $(E, t) \times F, s) = (E \cap F, t \times s)$, where $t \times s \in L^K$ is defined by $t \times s(k) = t(k) \otimes s(k)$,
   (d) $0 = (K, \underline{0}_L)$, $1 = (K, \underline{1}_L)$, where $\underline{\alpha}(k) = \alpha$ for arbitrarily $k \in K, \alpha \in L$.
2. The semiring $\mathcal{R}^* = (R, +^*, \times^*, 0^*, 1^*)$ is defined by
   (a) $(E, t), (F, s) \in R$, $(E, t) +^* (F, s) := (E \cap F, t \wedge s)$, where $t \wedge s$ is the infimum in L^K,
   (b) $(E, t), (F, s) \in R$, $(E, t) \times^* (F, s) = (E \cap F, t \oplus s)$, where $\oplus$ in L^K is defined component-wise.
   (c) $0^* = (K, \underline{1}_L)$, $1^* = (K, \underline{0}_L)$, where $\underline{\alpha}(k) = \alpha$ for arbitrarily $k \in K, \alpha \in L$,
3. $\neg : R \to R$ is defied by $(E, s) \in R$, $\neg(E, s) = (E, \neg s)$, where $\neg s$ is defined component-wise.

Then $\mathcal{F}(L) = (\mathcal{R}, \mathcal{R}^*)$ is the dual pair of semirings and the algebraic system of all L-fuzzy soft sets in X is isomorphic to the algebraic system of all $(\mathcal{R}, \mathcal{R}^*)$-fuzzy sets in X. ■

Example 2 Structure $(\mathcal{S}, \mathcal{S}^*)$ is defined by

1. The semiring $\mathcal{S} = (S, +, \times, \mathbf{0}, \mathbf{1})$ is defined by
   (a) $S = L^3$,
   (b) $(\alpha, \beta, \gamma) + (\alpha_1, \beta_1, \gamma_1) := (\alpha \vee \alpha_1, \beta \wedge \beta_1, \gamma \wedge \gamma_1)$,
   (c) $(\alpha, \beta, \gamma) \times (\alpha_1, \beta_1, \gamma_1) := (\alpha \otimes \alpha_1, \beta \wedge \beta_1, \gamma \oplus \gamma_1)$,
   (d) $\mathbf{0} = (0_L, 1_L, 1_L)$, $\quad \mathbf{1} = (1_L, 1_L, 0_L)$,
2. The semiring $\mathcal{S} = (S, +^*, \times^*, \mathbf{0}^*, \mathbf{1}^*)$ is defined by
   (a) $(\alpha, \beta, \gamma) +^* (\alpha_1, \beta_1, \gamma_1) := (\alpha \wedge \alpha_1, \beta \vee \beta_1, \gamma \vee \gamma_1)$,
   (b) $(\alpha, \beta, \gamma) \times^* (\alpha_1, \beta_1, \gamma_1) := (\alpha \oplus \alpha_1, \beta \vee \beta_1, \gamma \otimes \gamma_1)$,

(c) $\mathbf{0}^* = (1_L, 0_L, 0_L), \quad \mathbf{1}^* = (0_L, 0_L, 1_L)$,

Let $\neg : \mathcal{S} \to \mathcal{S}^*$ be defined by $(\alpha, \beta, \gamma) \in S, \quad \neg(\alpha, \beta, \gamma) = (\gamma, \neg\beta, \alpha)$.

Then $\mathcal{F}(L) = (\mathcal{S}, \mathcal{S}^*)$ is the dual pair of semirings. The algebraic structure $((\mathcal{S}, \mathcal{S}^*)^X, \sqcap, \sqcup, \neg, \leq)$ of $(\mathcal{S}, \mathcal{S}^*)$-fuzzy sets is isomorphic to the algebraic structure $(N(L), \cup, \cap, \neg, \leq)$ of neutrosophic L-fuzzy sets in a set X. ■

The basic tool in our process of unifying methods in fuzzy structures is the construction of two partially ordered monads defined by a dual pair of semirings. We repeat the definition of a partially ordered monad in the category **Set** of sets and mappings ([5, 8]).

Definition 5 $\mathbf{T} = (T, \Diamond, \eta, \ll)$ is an ordered monad in the category **Set**, if

1. $T : \mathbf{Set} \to \mathbf{Set}$ is a function defined on objects of **Set**,
2. $\ll = \{\leq_X : X \in \mathbf{Set}\}$, where $\leq_X$ is an order relation on the set $T(X)$,
3. For arbitrary objects $X, Y \in \mathbf{Set}$, on the set $Hom(X, T(Y))$ the order relation $\preceq_{X,Y}$ is defined such that for arbitrary mappings $h, k : X \to T(Y)$,

$$h \preceq_{X,Y} k \Leftrightarrow \forall x \in X, \quad h(x) \leq_Y k(x),$$

4. η is a system of mappings $\{\eta_X : X \to T(X) | X \in \mathbf{Set}\}$,
5. For each pair of mappings $f : X \to T(Y)$, $g : Y \to T(Z)$, there exists a composition (called a *Kleisli composition*) $g \Diamond f : X \to T(Z)$, which is associative,
6. For every mapping $f : X \to T(Y)$, $\eta_Y \Diamond f = f$ and $f \Diamond \eta_X = f$ hold,
7. $\Diamond$ is compatible with the composition of mappings, i.e. for mappings $f : X \to Y$, $g : Y \to T(Z)$, we have $g \Diamond (\eta_Y . f) = g.f$,
8. For arbitrary mappings $f, f' : X \to T(Y), g, g' : Y \to T(Z)$, the following implications hold.

$$g \preceq_{Y,Z} g' \Rightarrow g \Diamond f \preceq_{X,Z} g' \Diamond f,$$
$$f \preceq_{X,Y} f' \Rightarrow g \Diamond f \preceq_{X,Z} g \Diamond f'.$$

3 Closure Operators on Powerset Objects of $(\mathcal{R}, \mathcal{R}^*)$-Fuzzy Sets

In this section, we show how closure operators can be defined on powerset objects $(\mathcal{R}, \mathcal{R}^*)^X$ of all $(\mathcal{R}, \mathcal{R}^*)$-fuzzy sets in a set X. This general theory can subsequently be applied to any new L-fuzzy structure that can be transformed into a AMV-fuzzy set, i.e., L-fuzzy soft sets, L-neutrosophis fuzzy sets, L-intuitionistic fuzzy sets, etc., where L is a complete MV-algebra. The main tool for introducing this topological concept is a partially ordered monad, defined by a complete commutative idempotent semiring.

Theorem 1 *Let $\mathcal{R} = (R, +, \times, 0_R, 1_R)$ be the complete commutative idempotent semiring. The structure $\mathbf{T}_{\mathcal{R}} = (T, \Diamond, \eta, \ll_T)$ is defined by*

1. $T : \mathbf{Set} \to \mathbf{Set}$ *is such that* $T(X) = R^X$,
2. $\ll_T = \{\leq_X : X \in \mathbf{Set}\}$, *where for* $f, g \in T(X)$, $f \leq_X g \Leftrightarrow f(x)+g(x)=g(x)$ *for each* $x \in X$,
3. *For mappings* $f : X \to T(Y)$ *and* $g : Y \to T(Z)$ *we set*

$$g\Diamond f : X \to T(Z), \quad (g\Diamond f)(x)(z) = \sum_{y \in Y} f(x)(y) \times g(y)(z). \tag{1}$$

4. η_X *is the mapping* $X \to T(X)$ *defined by*

$$\eta_X(x)(y) = \begin{cases} 1_R, & x = y, \\ 0_R, & x \neq y, \end{cases}.$$

Then $\mathbf{T}_{\mathcal{R}}$ *is the ordered monad in the category* **Set**.

Remark 2 Let $(\mathcal{R}, \mathcal{R}^*)$ be a dual pair of semirings, where $\mathcal{R} = (R, +, \times, 0_R, 1_R)$, $\mathcal{R}^* = (R, +^*, \times^*, 0_R^*, 1_R^*)$. According to Theorem 1, there exist two ordered monads $\mathbf{T}_{\mathcal{R}} = (T, \Diamond, \eta, \ll_T)$ and $\mathbf{T}_{\mathcal{R}^*} = (T, \Diamond^*, \eta^*, \ll^*)$.

For arbitrary mapping $f : X \to Y$ in **Set** we can define two variants of extension maps $(\mathcal{R}, \mathcal{R}^*)^X \to (\mathcal{R}, \mathcal{R}^*)^Y$ monads $\mathbf{T}_{\mathcal{R}}$ and $\mathbf{T}_{\mathcal{R}^*}$, . In fact, two extensions $f^{\to}$ and $f^{\Rightarrow} : (\mathcal{R}, \mathcal{R}^*)^X \to (\mathcal{R}, \mathcal{R}^*)^Y$ can be defined by

$$s \in (\mathcal{R}, \mathcal{R}^*)^X, \quad f^{\to}(s) = (\eta_Y . f)\Diamond 1_{T(X)}, \quad f^{\Rightarrow}(s) = (\eta_Y^* . f)\Diamond^* 1_{T(X)}.$$

Using these two extension maps, we can define two categories of closure operators in powerset objects $(\mathcal{R}, \mathcal{R}^*)^X$.

Definition 6 Let $(\mathcal{R}, \mathcal{R}^*)$ be a dual pair of semirings and let $(\mathbf{T}_{\mathcal{R}}, \mathbf{T}_{\mathcal{R}^*})$ be a pair of monads defined by $(\mathcal{R}, \mathcal{R}^*)$. The category $\mathbf{Cl}(\mathcal{R})$ [$\mathbf{Cl}(\mathcal{R}^*)$, respectively] of closure operators in $(\mathcal{R}, \mathcal{R}^*)$-fuzzy sets is defined by

1. The objects of $\mathbf{Cl}(\mathcal{R})$ and $\mathbf{Cl}(\mathcal{R}^*)$ are (X, c), where
   (a) X is a set,
   (b) $c : T(X) \to T(X)$ is a closure operator in the ordered set $(T(X), \leq_X)$ [$(T(X), \leq_X^*)$, respectively], that is,
      i $\forall s, t \in T(X), \quad s \leq_X t \Rightarrow c(s) \leq_X c(t), \quad [s \leq_X^* t \Rightarrow c(s) \leq_X^* c(t)$, respectively],
      ii $\forall s \in T(X), \quad c(s) \geq_X s$, [$c(s) \geq_X^* s$, respectively],

iii $\forall s \in T(X), \quad cc(s) = c(s)$.

2. $f : (X, c_X) \to (Y, c_Y)$ is a morphism in $\mathbf{Cl}(\mathcal{R})$ [$\mathbf{Cl}(\mathcal{R}^*)$, respectively], if
   (a) $f : X \to Y$ is a mapping,
   (b) For each $s \in T(X)$ we have

$$f^{\to}(c_X(s)) \leq c_Y(f^{\to}(s)) \quad [f^{\Rightarrow}(c_X(s)) \leq_Y^* c_Y(f^{\Rightarrow}(s)), \text{ respectively}].$$

From the following proposition it follows that both categories $\mathbf{Cl}(\mathcal{R})$ and $\mathbf{Cl}(\mathcal{R}^*)$ are isomorphic.

Proposition 1 *There exists the isomorphic functor* $F : \mathbf{Cl}(\mathcal{R}) \to \mathbf{Cl}(\mathcal{R}^*)$.

Proof Instead of proof, we show only how the functor F is defined. If (X, c) is an object of $\mathbf{Cl}(\mathcal{R})$, we set $F(X, c) = (X, c')$, where the closure operator c' is defined by $c'(s) = \neg c(\neg s)$, where $\neg s \in R^X$ is defined by $(\neg s)(x) = \neg(s(x))$ and $\neg c$ is defined by $(\neg c)(s)(x) = \neg(c(x))$. For a morphism f we set $F(f) = f$. ■

From Theorem 1 it follows that $\mathbf{Cl}(\mathcal{R}$ represents the category of closure operators in R^X and $\mathbf{Cl}(\mathcal{R}^*)$ represents the category of interior operator in R^X.

As mentioned in the introduction, several new fuzzy structures can be transformed into $(\mathcal{R}, \mathcal{R}^*)$-fuzzy sets. All concepts and theories that apply to general $(\mathcal{R}, \mathcal{R}^*)$-structures can be applied to these fuzzy structures. Specifically, using Definition 6, we can define the notion of closure operators on the powersets of these new fuzzy structures and the continuous mappings between these closure operators. To illustrate it, we will show how to define these closure operators for $\mathcal{L}$-fuzzy soft sets using the transformation of $\mathcal{L}$-fuzzy soft sets into $(\mathcal{R}, \mathcal{R}^*)$-fuzzy sets from Example 1.

Example 3 Let $(\mathcal{R}, \mathcal{R}^*)$ be the dual pair of semirings from Example 1 and let $s \in R^X$ be a $(\mathcal{R}, \mathcal{R}^*)$-fuzzy set (ie, s can be identified with the L-fuzzy soft set in (K, X)). Using Example 1, for arbitrary $s \in R^X$ and $x \in X$, we have $s(x) := (E_x^s, \psi_x^s) \in R$, where $E_x^s \subseteq K$ and $\psi_x^s \in L^K$.

By this notation, we can present a method to illustrate how examples of closure operators on L-fuzzy soft sets can be constructed. Let X be a set, and let $d : 2^K \to 2^K$ and $g : L^K \to L^K$ be mappings such that

1. d is a closure operator on the ordered set $(2^K, \subseteq)$,
2. g is an extensional and idempotent operator on the ordered set $(L^K, \leq)$,
3. If $s, s' \in R^X$ are such that for arbitrary $k \in E_x^s \cap E_x^{s'}$, $\psi_x^s(k) \leq \psi_x^{s'}(k)$, we have $g(\psi_x^s)(k) \leq g(\psi_x^{s'})(k)$ for arbitrary $k \in E_x^s \cap E_x^{s'}$.

Let $c : R^X \to R^X$ be defined by

$$s \in R^X, x \in X, \quad c(s)(x) = (d(E_x^s), g(\psi_x^s)) \in R.$$

Then $(X, c) \in \mathbf{Cl}(\mathcal{R})$.

■

Example 4 Let $(\mathcal{S}, \mathcal{S}^*)$ be the dual pair of semirings from Example 2 and let X be a set. If $t \in S^X$, for $x \in X$ we set $t(x) = (t_1^x, t_2^x, t_3^x) \in S$. Let $e, i, j : L \to L$ be mappings such that

1. $e : (L, \leq) \to (L, \leq)$ is a closure operator,
2. $i, j : (L, \leq) \to (L, \leq)$ are interior operators.

Let $p : S^X \to S^X$ be defined by

$$t \in S^X, x \in X \in S, \quad p(t)(x) = (e(t_1^x), i(t_2^x), j(t_3^x)).$$

Then $(X, p) \in \mathbf{Cl}(\mathcal{S})$.

■

4 Conclusions and Future Development

The transformation of some new L-valued fuzzy structures into AMV-fuzzy sets $X \to F(L)$, where L is a complete MV-algebra, makes it possible to completely use existing theoretical results from the field of classical fuzzy sets for these new fuzzy structures as well. As an example of this approach, we show the possibility of introducing closure operators even for these new L-fuzzy structures, using monad theory, which is standardly used in classical fuzzy sets. It is understandable that this method also has its limitations. One of these limitations is that, in the presented form, this method can only be applied to fuzzy structures with values in the complete MV-algebra. The future development of this transformation method will therefore focus on the elimination of this requirement and subsequently on the application of this method to other theoretical tools used in the theory of fuzzy sets, such as topology on fuzzy structures.

References

1. Agarwal, M., Biswas, K.K., Hanmandlu, M.: Generalized intuitionistic fuzzy soft sets with applications in decision-making. Appl. Soft Comput. **13**(8), 3552–3566 (2013)
2. Atanassov, K.T.: Intuitionistic fuzzy sets. Fuzzy Sets Syst. **20**(1), 87–96 (1986)
3. Atanassov, K.T.: Intuitionistic fuzzy relations. In: Antonov, L. (ed.) III International School "Automation and Scientific Instrumentation", pp. 56–57. Varna (1984)
4. Babitha, K., John, S.: Hesitant fuzzy soft sets. J. New Results Sci. **2**(3), 98–107 (2013)
5. Eklund P., Galán M.A.: Partially ordered monads and rough sets. In: Peters, J.F., Skowron, A. (eds.) Transactions on Rough Sets VIII. Lecture Notes in Computer Science, vol. 5084. Springer, Berlin (2008)

6. Feng, F., Jun, Y.B., Zhao, X.Z.: Soft semirings. Comput. Math. Appli. **56**, 2621–2628 (2008)
7. Garg, H., Arora, H.: Generalized and group-based generalized intuitionistic fuzzy soft sets with applications in decision-making. Appl. Intell. **48**, 343–356 (2018)
8. Höhle, U.: Partially ordered monads. In: Many Valued Topology and Its Applications. Kluwer Academic Publishers, Boston, Dordrecht, London (2001)
9. James, J., Mathew, S.C.: Lattice valued neutrosophis sets. J. Math. Comput. Sci. **11**, 4695–4710 (2021)
10. Zhang, X., Bo, C., Smarandache, F., Dai, J.: New inclusion relation of neutrosophic sets with applications and related lattice structure. Int. J. Mach. Learn. Cybern. **9**, 1753–1763 (2018)
11. Maji, P.K., et al.: Fuzzy soft-sets. J. Fuzzy Math. **9**(3), 589–602 (2001)
12. Maji, P.K., Biswas, R., Roy, A.R.: Soft set theory. Comput. Math. Appl. **45**, 555–562 (2003)
13. Maji, P.K., et al.: An application of soft sets in a decision making problem. Comput. Math. Appl. **44**, 1077–083 (2002)
14. Molodtsov, D.: Soft set theory-First results. Comput. Math. Appl. **37**, 19–31 (1999)
15. Močkoř, J.: Semiring-valued fuzzy sets and F-transform. Mathematics **9**(23), 3107 (2021)
16. Močkoř, J., Hurtik, P., Hýnar, D.: Rough semiring-valued fuzzy sets with application. Mathematics (10), 2274 (2022). https://doi.org/10.3390/math10132274
17. Rodabaugh, S.E.: Relationship of algebraic theories to powerset theories and fuzzy topological theories for lattice-valued mathematics. Int. J. Math. Math. Sci. 1–71 (2007)
18. Solovyov, S.A.: Powerset operator foundations for catalg fuzzy set theories. Iran. J. Fuzzy Syst. **8**(2), 1–46 (2001)
19. Suo, C., Li, Y., Li, Z.: A series of information measures of hesitant fuzzy soft sets and their application in decision making. Soft Comput. **25**, 4771–4784 (2021)
20. Wang, J.-G., Li, X.-E., Chen, X.-H.: Hesitant fuzzy soft sets with application in multicriteria group decision making problems. Sci. World J. Article ID 806983 (2015)

An Application of a Fuzzy Multi-criteria Decision Making Process for Explainable Machine Learning in the Actuarial Context

Catalina Lozano-Murcia, Francisco P. Romero, Jesus Serrano-Guerrero, and Jose A. Olivas

1 Introduction

Data science has developed over the last decades a wide range of techniques focused on data preparation, prediction and classification by Artificial Intelligence (AI) models. In addition, Explainable Artificial Intelligence (XAI) defines methods and techniques to explain and understand the results or solutions proposed by AI models.

In general, AI model makers face the technical challenge of explaining black-box relationships and interpreting in simple language the results proposed by the solutions developed for the specific environment or context for which they were created and defining the reliability and consistency of their results. The development of the XAI allows different users to formalize the definition of interpretability and reliability of AI models, which are an intermediate objective to verify various criteria in the end-users of the results in the models made and informed decision making. One of the main areas of application of the XAI is the financial sector, especially insurance. This industry bases its profits on the correct capacity to manage future risks. This is fundamental to developing models that allow classifying (segmenting portfolios and commercialization), predicting (pricing and reserving) and transparent decision-making. The insurance industry represents an essential piece of the global economy when managing risks and public money, so its management is highly regulated. Therefore, non-replicable and non-auditable models are a problem, despite the advantages associated with efficiency in processes or their predictive capacity.

Another of the main applications that have been found to XAI in the insurance sector is associated with the prevention and detection of fraud. AI models and their

C. Lozano-Murcia · F. P. Romero (✉) · J. Serrano-Guerrero · J. A. Olivas
Department of Information Technologies and Systems, University of Castilla, La Mancha, Paseo de La Universidad, 4, 13071 Ciudad Real, Spain
e-mail: FranciscoP.Romero@uclm.es

M. E. Cornejo et al. (eds.), *Computational Intelligence and Mathematics for Tackling Complex Problems 5*, Studies in Computational Intelligence 1127,
https://doi.org/10.1007/978-3-031-46979-4_9

understanding have become powerful tools for detection and prevention, saving the insurance industry and its users millions of dollars annually. Many people consider that machine learning models generate non-intuitive or interpretable results and should not be used as a prediction or decision-making reference. Nevertheless, tools like Partial Dependence Plots (PDP) [1], SHAP [2], Decision Trees or scores such as Variable Importance (VIP) [3] allow identifying relationships and prioritizing determining variables, regardless of the objective of the model, nature of the variables or context. However, it is essential to note that these methods are not comparable, and their evaluations for different models do not yield directly analyzable results either. In this work, a fuzzy Multi-Criteria Decision Making approach is applied to the problem of pricing in health insurance. The analysis of the models with the best results using XAI techniques and the fuzzy methods to analyze these results allows for evaluating the relationships established by the models globally, considering an analysis of comparison of experts. The rest of the paper is organized as follows: Sect. 2 presents shortly the fuzzy MCDM approach. Section 3 shows a case study based on Labor Insurance claims where different techniques of ML, XAI and MCDM are applied. Finally, conclusions and future work will be commented on in Sect. 4.

2 Approach

This paper addresses the problem of identifying variables that may affect a machine learning model's decision (segmentation, pricing, forecasting) in an actuary problem. We use XAI techniques to rank variables according to their relevance in decision-making. After applying fuzzy methods for consensus modelling and uncertainty management, a new ranking of variables can be obtained according to their significance, both from a global and local point of view. The proposed approach is summarized in Fig. 1, and its main components are described as follows::

- Actuary Problem Definition of a problem in the actuarial context.
- ML Problem: Formulation of the problem to get a machine learning based solution.
- ML Approach: Application of several machine learning models (RF, Boosting, ANN, etc.) to solve the machine learning problem.
- Evaluation and Filter: Evaluate the solutions of the model according to the selected metrics and filter those models with enough accuracy.
- Global explanations: Explanations about the feature relevance of all predictions. Techniques such as VIP, Decision Trees, Model Explanation are used.
- Local Explanations: Explanations about the feature relevance of local results using methods like LIME or SHAP.
- MCDM: Fuzzy multicriteria decision making process to select those features more relevant to the problem.

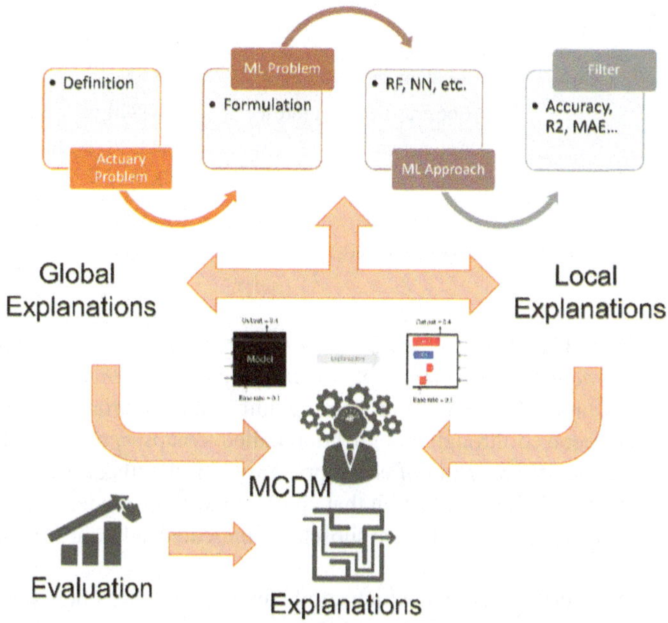

Fig. 1 Approach

3 Study Case

The main purpose of this problem is to predict surcharges related to a health insurance base [4]. The original basis has 1338 observations. It has 6 variables in addition to the target variable. Of the explanatory variables, 3 categorical and 3 numerical variables were identified, of which 3 observations were eliminated due to inconsistencies in the value of the objective variable: Age of the insured, Gender of the insured. (female or male), Body mass index (bmi), Number of children, Smoker (yes) or non-smoker (no), place of residence: northeast, northwest, southeast, southwest.

3.1 Models

To determine some decision alternatives, we apply 4 kind of models:

- Regression Trees: Establishes monotonous transformations of independent predictive variables with a determination a recursive algorithm, divided into hyper rectangles where each observation contained will have same estimated value. (CHAID—Chi Square Automatic Interaction Detector, QUEST—Quick Unblased Efficient Statistical Tree, CART—Classification and Regression Tree) [5].

Table 1 R^2 by model

Model	R^2	MAE
Random decision forest	0.85096	2585.0
Boostrap decision forest	0.84890	2540.6
Autonetwork	0.82226	2372.3
Auto structure suggestion	0.81043	2741.9
Decision tree	0.75216	3247.0
Linear regression	0.74460	4275.42

- Artificial Neural Networks: *A neural network is a computational, parallel model, composed of adaptive processing units with a high interconnection in them* [6]. These computational model use processing units called neurons that process the information received through a connection called synapses, all in hidden layers that finally generate an output of either prediction or classification.
- Bagging [7]: Bootstrap aggregation that combines various models from an initial family of models that train different subsets, as indicated in figure 5 of data whose process seeks to reduce variance.
- Boosting [8]: Additive method which combines a variety of models since it seeks through the general use of models of little effectiveness the determination of a better predictor by reducing the prediction error in the process.

More than 30 models were executed based on data without any transformation, seeking to evaluate the discrimination capacity of different techniques. From this exercise, 5 models with R^2 higher than 75% were obtained, which are presented in Table 1, also Linear regression model result. To contrast and explain the relationships established in these models, an XAI analysis was carried out, which is explained below.

3.2 *XAI*

To identify the relationships established and the incidence of the variables in each one, an explainability approach was proposed through 2 techniques:

- VIP (Variable Importance) is a method that allows the determination of each variable's global contribution in the prediction model. Determination methods may vary by model, which does not make scores comparable. However, the order of the variables allows us to evaluate the global relevance and compare the established relationships.
- Decision Tree: To identify the relationships that a model establishes, a common technique is to train decision trees with the input variables and their respective outputs from the model, thus the tree identifies and classifies the relationships that lead to the estimated result.

Table 2 Variable importance by model

Model	Children (%)	Field1 (%)	Age (%)	Bmi (%)	Smoker.no (%)	Region southeast (%)	Sex female (%)
Random decision forest	1.25	2.81	11.37	14.63	68.27		
Boostrap decision forest	1.19	1.78	11.45	17.88	66.96		
Autonetwork			16.14	12.82	51.73	3.99	4.31
Auto structure suggestion	3.31		17.21	12.04	56.23		3.26
Decision tree	3.40	10.53	16.38	24.56	39.94		

The Table 2 reflects different results for the models obtained. However, the relevance of smoker.no, as the most important in all models, and BMI and age that dispute the according and third place in several of the models. Given these differences, a complementary analysis of Decision Support Results was carried out, which is presented below.

3.3 *Decision Making Process*

A Decision-Making process is used to contrast the established relationships and evaluate the general influence of the variables on the problem posed. These systems allow to direct the decision-making process in an organized way, contrasting results of different experts to carry out an evaluation on a set of possible alternatives [8]. These techniques have been becoming popular in the financial sector thanks to Fintech and Business Intelligence (BI) [9]. For the case study the experts are the results of the 5 models with the best results of R^2. To establish the analysis, the percentages VI average of each variable were found, obtaining, in a generalized way, a new scale that considers the information of each model and its weight of predictive capacity, obtaining the results presented in Table 3.

Given the results of relevance of variables the process would suggest that be-yond the best R^2 one should consider that model that weighs the same general importance of the variables, a general requirement that the Random decision forest model fulfills.

4 Conclusions

XAI techniques were used to overcome the interpretability problem of the five black box models obtained and to facilitate understanding of the relationships established by the models and their intuitive contrast depending on the context of the problem.

Table 3 VI average

Variable	VI average (%)
Smoker.no	56.6
Bmi	16.4
Age	14.5
Field1	5.0
Children	2.3
Sex female	1.5
Region southeast	0.8

Finally, by using a Decision-Making process it was possible to contrast the general relevance of the variables in the problem and base the selection decision not only on the criterion of predictive capacity through the contrast of R^2 or MAE if not also on the representativeness of the weight of the variables. A future is incorporating more sophisticated XAI and Decision Support Systems techniques, which allow automating the process of understanding and selection of models massively and practically for the developer.

References

1. Greenwell, B.M.: PDP: an R package for constructing partial dependence plots. R J. **9**(1), 421 (2017)
2. Lundberg, S.M., Lee, S. I.: A unified approach to interpreting model predictions. Adv. Neural Inf. Proc. Syst. **30** (2017)
3. Greenwell, B.M., Boehmke, B.C., Gray, B.: Variable importance plots-an introduction to the vip package. R J. **12**(1), 343 (2020)
4. Lantz, B.: Machine Learning with R: Expert Techniques for Predictive Modeling. Packt Publishing Ltd. (2019)
5. Hassoun, M.H.: Fundamentals of Artificial Neural Networks. MIT Press (1995)
6. Krasheninnikova, E., García, J., Maestre, R., Fernández, F.: Reinforcement learning for pricing strategy optimization in the insurance industry. Eng. Appl. Artif. Intell. **80**, 8–19 (2019)
7. Henckaerts, R., Côté, M.P., Antonio, K., Verbelen, R.: Boosting insights in insurance tariff plans with tree-based machine learning methods. North Am. Actuar. J. **25**(2), 255–285 (2021)
8. Wibowo, S., Grandhi, L., Grandhi, S., Wells, M.: A fuzzy multicriteria group decision making approach for evaluating and selecting fintech projects. Mathematics **10**(2), 225 (2022)
9. Belle, V., Papantonis, I.: Principles and practice of explainable machine learning. Front. Big Data **39** (2021)

A Software Tool for Solving the Traveling Salesman Problem and Related Non-fuzzy and Fuzzy Optimization Problems

Boldizsár Tüű-Szabó, Ruba Almahasneh, Péter Földesi, and László T. Kóczy

1 Introduction

The Traveling Salesman Problem (TSP) is a well-known combinatorial optimization problem. The goal is to find the shortest tour that visits each city in a given list exactly once and then returns to the starting point. The problem is easy to describe, but hard to solve, it belongs to the NP-hard class. The importance of the TSP arises not only from its theoretical appeal but also from the variety of its applications such as logistics, chip manufacturing, genetics and computer wiring. However the original problem has also many application areas as mentioned, but in some areas especially in logistics and transportation more complicated models are required. These extended versions of the original TSP problem also belong to the NP-hard class. With our software the TSP and some extended versions can be solved. In the software the DBMEA was implemented which is an efficient method for solving the TSP and its variants resulting in optimal or near-optimal solutions within reasonable amount of time [1–3]. However many evolutionary computation frameworks were presented in recent years (DEAP Python [4], Evolving Objects (EO) C++ framework [5] etc.), but what they have in common is that they only focus on the most widely used evolutionary algorithms (such as Genetic Algorithm, Genetic Programming,

B. Tüű-Szabó (✉) · L. T. Kóczy
Department of Information Technology, Széchenyi István University, Györ, Hungary
e-mail: tuu.szabo.boldizsar@sze.hu

R. Almahasneh
Telecommunications and Media Informatics, Budapest University of Technology and Economics, Budapest, Hungary

P. Földesi
Department of Logistics, Széchenyi István University, Györ, Hungary

M. E. Cornejo et al. (eds.), *Computational Intelligence and Mathematics for Tackling Complex Problems 5*, Studies in Computational Intelligence 1127,
https://doi.org/10.1007/978-3-031-46979-4_10

Evolutionary Strategies, and Particle Swarm Optimization) and its operators. Another drawback is that these frameworks don't contain local search methods preventing the design of memetic algorithms.

2 The Discrete Bacterial Memetic Evolutionary Algorithm

The DBMEA is the memetic version of the Bacterial Evolutionary Algorithm [6], it combines the discrete version of the BEA with local search techniques (2-opt and 3-opt). Recently the memetic algorithms are very popular in the field of optimization because it can be efficient in solving NP-hard optimization problems. This combination can eliminate the weaknesses of both methods. Evolutionary algorithms search in the global search which often cause slow convergence speed. Local search methods search in the neighbourhood of the current solution, so they often fail to find the global optimum.

The process of the DBMEA consists of the following steps:

- Creating the Initial Population

In each iteration until the stoppage criterion is met:

- Bacterial mutation for each bacterium
- 2-opt and 3-opt local search
- Gene transfer performed on the population

In DBMEA the bacteria form a population. Each bacterium represents a possible solution for the examined problem. In DBMEA the bacteria are coded with a very simple permutation encoding. The starting point of the tour is indexed with 0 (for each tour it is the starting node, so it does not appear in the codes), and other nodes are mapped with a unique index (1…n). The tours are expressed as the sequence of these indices clearly identifying the tour. Each node needs to visit once, so the length of the tours is n − 1.

The bacterial mutation operation works on the bacterium individually resulting that each bacterium in the population becomes more or in worst case equally fit (see details in [1]). In our algorithm two different types of mutation, the coherent segment mutation (segments consist of adjacent elements) and the loose segment mutation (the members of a segment are not necessarily adjacent) can be selected.

In DBMEA the local search combines the 2-opt and 3-opt techniques. First, the tour is improved by 2-opt steps. And when no further improvement is possible with 2-opt, then 3-opt will be applied for the tour. The local search is stopped, when the tour cannot be improved further by 3-opt steps. 2-opt local search replaces two edge pairs in the original graph to reduce the cost of the tour. The edge exchange is continued till no further improvement is possible. The 3-opt local search improves the tour by replacing three edges with three other. There are four ways to reconnect the tour in 3-opt. The output of the 3-opt step is always the less costly tour.

The gene transfer operation provides the information transfer within the population promoting the development of the bacteria (see details in [1]).

3 The Software

The following NP-hard discrete optimization problems can be solved with the proposed software:

- The Traveling Salesman Problem: It is one of the most studied discrete optimization problems because of its diverse range of applications such as mathematics, computer science, engineering, logistics and genetics. The aim is to find the shortest possible route that visits every city exactly once and returns to the starting point [7].
- The Traveling Salesman Problem with Time Windows (TSPTW): In this problem each customer has a service time and a time window defining its ready time and due time. The task is to find the minimum cost tour, which satisfies every time windows [2].
- The Traveling Repairman Problem (TRP): It is a cumulated version of the original TSP problem. The task is to find a Hamiltonian circuit that minimizes the sum of arrival times at each node [3].
- The Time Dependent Traveling Salesman Problem (TDTSP): It was proposed by Schneider [8]. It this problem a traffic jam region and the rush hours are defined. In the rush hours the edge lengths between two nodes in the traffic region are multiplied with a predefined jam factor.
- The One-Commodity Pickup-and-Delivery Traveling Salesman Problem (1-PDTSP): It was introduced by Hernández-Pérez in 2004 [9]. One kind of products is transported from some nodes to others. The customers are divided into two groups: the delivery (they supply a given amount of product) and pickup customers (they demand a given amount of product). The value of demanding or delivering amount is assigned to each node. The products are transported with one vehicle which has a maximum capacity. The aim is to find a feasible solution with lowest cost which visits each node only once starting from the depot.
- Fuzzy modifications of the TDTSP model: The TDTSP model contains simplifications (only one rush hours period was defined, the jam factor value is constant within the rush hours period, jam factor value doesn't vary within the jam region), so its applicability is limited. The 3FTDTSP model eliminates the drawbacks of the original model [10]:
  - Triangular fuzzy costs for the distances between the edges
  - Membership function(s) of the fuzzy jam region(s) J providing more realistic description of the traffic conditions
  - Membership function(s) of the traffic rush hour time period(s) R

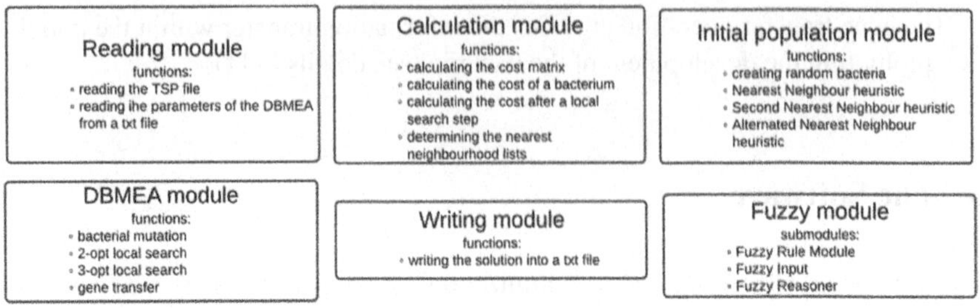

Fig. 1 The modules of the software

Another more realistic model was developed using interval values type 2 fuzzy sets [11] for the description of the rush hour periods and the jam regions.

The software was developed with C++. The software was divided into the following modules (Fig. 1).

- Reading module: Before the process of the DBMEA the read of the instance file in TSPLIB format and the parameters of the algorithm are needed. The parameters are read from a txt file which also contains the name of the optimization problem. For solving the TDTSP the four coordinates of the traffic jam region, the speed and the rush hours period must also be given.
- Calculation module: This module contains all of the functions which contain calculation operations. The TSP file contains the coordinates of the vertices therefore the calculation of cost matrix is needed. The examined optimization problems differ in the cost functions which are implemented in this module. For TSPTW and 1-PDTSP the feasibility checking of the tours is also needed. A local search steps mean deleting some edges from the solution and adding other edges resulting in a valid tour. While for TSP the calculation of the tour after a local search step is trivial for the other problems it is more complicated. The calculation module contains the functions which calculate the cost value of the tour after a local search step for each optimization problem. The local search can be accelerated with only taking into account the elements of the nearest neighbour lists in the local search steps.
- Initial population module: In this module the random creation of the bacteria and three tour construction heuristics were implemented: Nearest Neighbour Heuristic which constructs a tour in which the nearest unvisited city is always visited), Second Nearest Neighbour Heuristic which always visits the second nearest unvisited city in the tour), Alternated Nearest Neighbour Heuristic which combines the two methods, the nearest and second nearest unvisited cities are visited in alternating order. In the future this module will be extended with other promising tour construction heuristics.
- DBMEA module: It contains the main operators of the algorithm (bacterial mutation, 2-opt and 3-opt local search, gene transfer operation). For the detailed process of this operators see II. Section. Each operator contains cost calculation steps; the appropriate cost function is called from the calculation module.

- Writing module: Finally the best bacterium in the population will be the solution. This solution (the visiting order of the vertices) with its cost and runtime value will be written into a txt file.
- Fuzzy module: The software was built to handle type1 and type 2 fuzzy sets, divided into the following sub modules:

1. *Type 1 Fuzzy Sets*:
    - Fuzzy Rule Module: This module consists of the linguistic variables and fuzzy rules. This class main purpose to translate and sort the fuzzy rules provided for the case under study.
    - Fuzzy Input: This class is responsible of handling the input variables for any case under study.
    - Fuzzy Reasoner: This class is responsible of calculating the fuzzy rule output values and it performs the final defuzzification process. Basically, it takes the inputs and the fuzzy rules provided, then each fuzzy rule is mapped to output variable and lastly it performs the defuzzification process. In the 3FTDTSP, we used the center of gravity (COG) [12], which is one of the most commonly used defuzzification methods (other defuzzification methods are; mean of maximum (MOM) [13–15], and center average methods, the center of area method (COA) [15]). Furthermore, The weighted average method [12] for finding the crisp output value, y^* is computed by taking the sum of the multiplication of each weighting function y, with the maximum value of its respective membership value y, and dividing it by the sum of the weighting functions. This representation was chosen while implementing the 3FTD TSP C++ program due to computational complexity (calculation of COA and weighted average method show very close results). Finally, these crisp numbers y^* are summed up providing the total distance of the tour.
2. *Type 2 Fuzzy Sets*: The Type 2 fuzzy sets module is complementary of the first module where additional calculation must be performed to perform the intuitionistic part. Interval-valued intuitionistic fuzzy sets were used for modelling the uncertainties (for details see [16]).

4 Conclusions

Our aim was to develop a software tool for solving the TSP and closely related optimization problems. The DBMEA was implemented in the software because our experiments show that it is very efficient method for solving the TSP and its extensions. In the future we would like to extend the software with other popular and efficient evolutionary algorithms (Genetic Algorithm [17], Ant Colony Optimization [18], Bee Colony Optimization [19], Particle Swarm Optimization [20] etc.). Building on the already implemented modules our software will provide an opportunity for fair comparison of the different evolutionary algorithms's performances for solving the examined optimization problems.

Acknowledgements The authors would like to thanks to the Hungarian Scientific Research Fund (OTKA) K124055 for the support of the research.

References

1. Kóczy, L.T., Földesi, P., Tüű-Szabó, B.: An effective discrete bacterial memetic evolu-tionary algorithm for the traveling salesman problem. Int. J. Intell. Syst. **32**(8), 862–876 (2017)
2. Kóczy, L.T., Földesi, P., Tüű-Szabó, B.: Enhanced discrete bacterial memetic evolu-tionary algorithm-an efficacious metaheuristic for the traveling salesman optimization. Inf. Sci. **460–461**, 389–400 (2018)
3. Tüű-Szabó, B., Földesi, P., Kóczy, L.T.: An efficient evolutionary metaheuristic for the traveling repairman (minimum latency) problem. Int. J. Comput. Intell. Syst. **13**(1), 781–793 (2020)
4. Félix-Antoine, F., De Rainville, F.-M., Gardner, M.-A., Gagné, C., Parizeau, M.: DEAP: evolutionary algorithms made easy. J. Mach. Learn. Res. **13**, 2171–2175 (2012)
5. Keijzer, M., Merelo, J.J., Romero, G., Schoenauer, M.: Evolving Objects: A general purpose evolutionary computation library. In: Artificial Evolution. EA (2001). Lecture Notes in Computer Science, vol. 2310. Springer, Berlin (2001)
6. Nawa, N.E., Furuhashi, T.: Fuzzy system parameters discovery by bacterial evolution-ary algorithm. IEEE Trans. Fuzzy Syst. **7**, 608–616 (1999)
7. Applegate, D.L., Bixby, R.E., Chvátal, V., Cook, W.J.: The Traveling Salesman Problem: A Computational Study, pp. 1–81. Princeton University Press, Princeton (2006)
8. Schneider, J.: The time-dependent traveling salesman problem. Phys. A **314**, 151–155 (2002)
9. Hernández-Pérez, H., Salazar-González, J.J.: Heuristics for the one-commodity pickup and-delivery travelling salesman problem. Transp. Sci. **38**, 245–255 (2004)
10. Kóczy, L.T., Földesi, P., Tüű-Szabó, B., Almahasneh, R.: Modeling of Fuzzy Rule-base Algorithm for the Time Dependent Traveling Salesman Problem, IEEE International Con-ference on Fuzzy Systems (FUZZ-IEEE), 23–26 June 2019. LA, USA, New Orleans (2019)
11. Atanassov, R.K.: Intuitionistic fuzzy sets. Fuzzy Sets Syst. **20**, 87–96 (1986)
12. Voskoglou, M.G.: Comparison of the COG defuzzification technique and its variations to the GPA index. Am. J. Comput. Appl. Math. **6**(5), 187–193 (2016)
13. Berenji, H.R.: Fuzzy logic controllers. In: Yager, R.R., Zadeh, L.A. (Eds.) An Introduction to Fuzzy Logic Applications in Intelligent Systems. The Springer International Series in Engineering and Computer Science (Knowledge Representation, Learning and Expert Sys-tems), vol. 165. Springer, Boston, MA (1992)
14. Patel, A.V., Mohan, B.M:: Some numerical aspects of center of area defuzzification method. Fuzzy Sets Syst. **132**(3), 401–409 (2002)
15. Chakraverty, S., Sahoo, D.M., Mahato, N.R.: Defuzzification. In: Concepts of Soft Computing. Springer, Singapore (2019)
16. Almahasneh, R., Tüű-Szabó, B., Kóczy, L.T., Földesi, P.: Optimization of the time-dependent traveling salesman problem using interval-valued intuitionistic fuzzy sets. Axioms **9**, 53 (2022)
17. Holland, J.H.: Adaption in Natural and Artificial Systems. The MIT Press, Cambridge (1982)
18. Dorigo, M., Birattari, M., Stutzle, T.: Ant colony optimization. IEEE Comput. Intell. Mag. **1**(4), 28–39 (2006)
19. Lučić, P., Teodorović, D.: Transportation modeling: an artificial life approach. In: Proceedings of the 14th IEEE International Conference on Tools with Artificial Intelligence, pp. 216–223. Washington, DC (2002)
20. Kennedy, J., Eberhart, R.: Particle swarm optimization. In: Proceedings of the IEEE International Conference on Neural Networks ICNN 1995, Perth, WA, Australia, vol. 4, pp. 1942–1948 (1995)

Characterizing Clarify and Reduce Method by Means of Left-Sided Formal Concept Analysis

M. José Benítez-Caballero and Jesús Medina

1 Introduction

There are diverse mathematical tools to manage information collected in databases, such as, Formal Concept Analysis (FCA) [4]. Attribute reduction is a relevant problem in all of these tools, which philosophy is different depending on the considered tool. In FCA, the main idea is to obtain minimal sets of attributes which concept lattice is isomorphic to the original one, which are called reducts. In the case of multi-adjoint formal concept analysis [6], the consideration of the join-irreducible elements is essential to compute the reducts [2]. On the other hand, the Clafify-and-Reduce method (CR-method) consists of performing a few steps. Clarify the formal context consist of unifying those columns that are equal. After that, the columns that can be expressed as the infimum of other columns [5] are also removed.

In this paper, we focus on the relation between both reduction mechanisms, the one introduced in multi-adjoint formal concept analysis and the CR-method, and we will show that the last one is a particular case of the former.

Partially supported by the 2014–2020 ERDF Operational Programme in collaboration with the State Research Agency (AEI) in projects PID2019-108991GB-I00 and PID2022-137620NB-I00, and with the Department of Economy, Knowledge, Business and University of the Regional Government of Andalusia in project FEDER-UCA18-108612, and by the European Cooperation in Science & Technology (COST) Action CA17124.

M. J. Benítez-Caballero (✉) · J. Medina
Departament of Mathematics, Universidad de Cádiz (Puerto Real), Cadiz, Spain
e-mail: mariajose.benitez@uca.es

J. Medina
e-mail: jesus.medina@uca.es

M. E. Cornejo et al. (eds.), *Computational Intelligence and Mathematics for Tackling Complex Problems 5*, Studies in Computational Intelligence 1127,
https://doi.org/10.1007/978-3-031-46979-4_11

2 Preliminaries

In this section, several previous notions and results are recalled.

2.1 Left-Sided Formal Concept Analysis

The main adjoint triple [3] considered in this paper is the left-sided adjoint triple. Given two posets $(P, \leq_P, \top_P, \bot_P)$ and $(Q, \leq_Q, \top_Q, \bot_Q)$, the triple $(\&_l, \nwarrow_l, \swarrow^l)$, where $\&_l : P \times Q \to P$, $\nwarrow_l : P \times P \to Q$, and $\swarrow^l : P \times Q \to P$ are defined, for all $x, z \in P$, $y \in Q$, as

$$z \nwarrow_l x = \begin{cases} \top_Q & \text{if } x \leq_P z \\ \bot_Q & \text{otherwise} \end{cases} \quad z \swarrow^l y = \begin{cases} \top_P & \text{if } y = \bot_Q \\ z & \text{if } y \neq \bot_Q \end{cases} \quad x \&_l y = \begin{cases} x & \text{if } y \neq \bot_Q \\ \bot_P & \text{if } y = \bot_Q \end{cases}$$

is an adjoint triple [3], which is called *left-sided adjoint triple*.

Clearly, the conjunctor $\&_l$ fulfills the boundary condition with the top element on the second argument, that is, $x \& \top_Q = x$, for all $x \in P$. From now on, we will consider an adjoint formal context $(\mathcal{O}, \mathcal{P}, \mathcal{R})$, where $\mathcal{O}$ is the set of the objects, $\mathcal{P}$ is the set of attributes and $\mathcal{R}$ is the relation between them; and the left-sided adjoint frame $(P, Q, P, \&_l, \nwarrow_l, \swarrow^l)$.

In order to define the concept-forming operators, we will consider the crisp set of objects $Q^{\mathcal{O}}$ and set of fuzzy subset of attributes $P^{\mathcal{P}}$, with $f : \mathcal{P} \to P$. Therefore, the concept-forming operators, ${}^{\uparrow_l} : 2^{\mathcal{O}} \longrightarrow P^{\mathcal{P}}$ and ${}^{\downarrow^l} : P^{\mathcal{P}} \longrightarrow 2^{\mathcal{O}}$, are defined, for all $X \in 2^{\mathcal{O}}$, $f \in P^{\mathcal{P}}$ and $a \in \mathcal{P}$, $x \in \mathcal{O}$, as:

$$X^{\uparrow_l}(a) = \inf\{\mathcal{R}(x, a) \swarrow^l X(x) \mid x \in \mathcal{O}\} \tag{1}$$

$$f^{\downarrow^l}(x) = \inf\{\mathcal{R}(x, a) \nwarrow_l f(a) \mid a \in \mathcal{P}\} \tag{2}$$

These concept-forming operators form a Galois connection [6]. As usual, a *left-sided concept* is a pair $\langle g, f \rangle$ verifying that $X^{\uparrow_l} = f$ and $f^{\downarrow^l} = X$, with $X \in 2^{\mathcal{O}}$, $f \in P^{\mathcal{P}}$. Moreover, the set X is named *extent* and the set f is denominated *intent*. Indeed, the extent will be denoted as Ext_l and the intent as Int_l.

The following definition rewrites the notion of concept lattice when the left-sided adjoint triple is considered.

Definition 1 The *left-sided concept lattice* associated with an adjoint frame $(P, Q, P, \&_l, \nwarrow_l, \swarrow^l)$ and a context $(\mathcal{O}, \mathcal{P}, \mathcal{R})$ is the set

$$\mathcal{M}_l(\mathcal{O}, \mathcal{P}, \mathcal{R}) = \{\langle X, f \rangle \mid X \subseteq \mathcal{O}, f \in P^{\mathcal{P}} \text{ and } X^{\uparrow_l} = f, f^{\downarrow^l} = X\}$$

in which the ordering is defined by $\langle X, f_1 \rangle \preceq_l \langle X', f_2 \rangle$ if and only if $X \subseteq X'$ (equivalently $f_2 \preceq_1 f_1$). The associated concept lattice is denoted as $(\mathcal{M}_l, \preceq_l)$.

2.2 The Clarify-and-Reduce Method

The clarification and reduction of a formal context [4] can be applied to the set of objects or the set of attributes -the one considered in this paper- and it can be used in several FCA approaches, such as, in the multi-adjoint framework [6]. In order to recall these notions, we will consider a context $(\mathcal{O}, \mathcal{P}, \mathcal{R})$ and the derivation operators $^{\downarrow}$, $^{\uparrow}$ in the classical case [4].

This method has several steps. The first one is to clarify the formal context which consists in unifying the duplicated columns. As a consequence, the new attributes will be defined as the classes $[a] = \{a_i \in \mathcal{P} \mid a_i^{\downarrow} = a^{\downarrow}\}$, that is, $\mathcal{P}^{\bullet} = \{[a] \mid a \in \mathcal{P}\}$. Therefore, a clarified formal context will be denoted as $(\mathcal{O}, \mathcal{P}^{\bullet}, \mathcal{R}^{\bullet})$, where the relation $\mathcal{R}^{\bullet}$ is defined as $\mathcal{R}^{\bullet}(x, [a]) = 1$ if and only if $\mathcal{R}(x, a_i) = 1$, with $a_i \in [a]$, for any object $x \in \mathcal{O}$ and class attribute $[a] \in \mathcal{P}^{\bullet}$.

After that, we reduce the clarified formal context. A clarified and reduced formal context will be denoted, in this paper, as CR-formal context. The set of extents will be denoted as $\mathrm{Ext}_{\mathrm{CR}}$ and the set of intents as $\mathrm{Int}_{\mathrm{CR}}$. In this step, we eliminate the reducible attributes from the formal context. A *reducible attribute* is an attribute $[a] \in \mathcal{P}^{\bullet}$ such that the concept $\langle [a]^{\downarrow}, [a]^{\downarrow\uparrow} \rangle$ is a meet-reducible concept in the clarified context $(\mathcal{O}, \mathcal{P}^{\bullet}, \mathcal{R}^{\bullet})$.

This method is denominated as *CR-method* in [5]. Furthermore, in the aforementioned paper, CR-method is adapted to several extensions of FCA, as the three-way concept lattices [7], the variable threshold concept lattices [9] and the generalized one-sided concept lattices [8], with the following structure:

- **Step 1: Scaling**. First of all, several scale attributes will be considered instead of the original ones. This step can be trivial in some FCA frames, as in three-way concept lattices or attribute-induced three-ways concept lattices.
- **Step 2: Clarification and reduction**. Once the formal context is scaled, we use the ideas formerly mentioned in order to clarify and reduce the scaled formal context.
- **Step 3: Discernibiltiy function**. Now, we are able to use the attributes of the scaled CR-formal context in order to build a conjunctive normal form (CNF), named *discernibility function*:

$$\bigwedge_{[a] \in \mathcal{P}^{\bullet}} \left(\bigvee_{a_i \in [a]} a_i \right)$$

- **Step 3: Replacement**. The final step is to replace the scaled attributes by the original ones. Therefore, the reducts of the original formal context are defined as the clause of the reduced disjunctive normal form (DNF) of the discernibility function.

3 Generalizing CR-Method

In this section, we are going to present some results connecting the CR-method and the reduction obtained with the multi-adjoint philosophy considering the left-sided adjoint triple on the unit interval. The first result asserts that the extents in both environment are the same.

Theorem 1 *Considering a formal context* $(\mathcal{O}, \mathcal{P}, \mathcal{R})$, *we have that*

$$Ext_{CR}(\mathcal{O}, \mathcal{P}^{\bullet}, \mathcal{R}^{\bullet}) = Ext_l(\mathcal{O}, \mathcal{P}, \mathcal{R})$$

As a consequence, the concept lattices are isomorphic. Moreover, the scaled values of the attributes defining a column in the clarified and reduced formal context define a threshold over the value taken for that attribute when the left-sided philosophy is considered.

Proposition 1 *Let* $y = \{a_{1\geq\delta_{a_1}}, a_{2\geq\delta_{a_2}}, \ldots, a_{n\geq\delta_{a_n}}\}$ *be the set of all of the attributes associated with a column in the scaled CR-formal context. For any subset of objects* $X \in \mathcal{O}$, *we have that if* $y \in X^{\uparrow}$, *then* $X^{\uparrow_l}(a_i) \geq \delta_{a_i}$, *for all* $i \in \{1, \ldots, n\}$.

Moreover, for any subset of objects, the values taken by the attributes in its intent are related to the scaled values for that attributes when the CR philosophy is used.

Proposition 2 *Given a subset of objects* $X \subseteq \mathcal{O}$, *we have that* $X^{\uparrow} = \{y_1, y_2, \ldots, y_k\}$ *if and only if* $X^{\uparrow_l}(a) = \max\{\delta \mid a_{\geq\delta} \in y_j, j \in \{1, 2, \ldots, k\}\}$

As a consequence, the reducts obtained from both theories are the same.

Theorem 2 *Given* $D \in \mathcal{P}$, *we have that* D *is a reduct for the CR-formal context if and only if* D *is a reduct for the left-sided formal context.*

We will detail the results exposed through an example in the sequel.

Example 1 The variable threshold formal concept described in [5], adding a new attribute, will be considered in this example. We have the formal context $(\mathcal{O}, \mathcal{P}, \mathcal{R})$, where four objects are considered $\mathcal{O} = \{x_1, x_2, x_3, x_4\}$, and six observed features $\mathcal{P} = \{a, b, c, d, e, f\}$ are in the set of attributes. The relationship between them is shown in Fig. 1.

First of all, we are going to study the formal context with the CR-method philosophy. We will scale the formal context, as the initial step of the CR-method. After that, we must clarify and reduce the formal context. We obtain the CR-formal context, presented in Table 1.

Now, we are able to create the discernibility function using the attributes of the CR-formal context, which is described as follows:

$$(a_{\geq 0.6} \vee c_{\geq 0.9} \vee e_{\geq 0.9}) \wedge (a_{\geq 1.0} \vee b_{\geq 0.9}) \wedge (c_{\geq 1.0} \vee e_{\geq 1.0}) \wedge d_{\geq 0.5}$$

$\mathcal{R}$	a	b	c	d	e	f
x_1	0.5	0.7	0.7	0.5	0.7	0.7
x_2	0.6	0.7	1.0	0.5	1.0	0.7
x_3	1.0	0.9	1.0	0.1	1.0	1.0
x_4	1.0	0.9	0.9	0.1	0.9	0.7

$\mathcal{R}^\bullet$	$\{a_{\geq 0.6}, c_{\geq 0.9}, e_{\geq 0.9}\}$	$\{a_{\geq 1.0}, b_{\geq 0.9}\}$	$\{c_{\geq 1.0}, e_{\geq 1.0}\}$	$\{d_{\geq 0.5}\}$
x_1	0	0	0	1
x_2	1	0	1	1
x_3	1	1	1	0
x_4	1	1	0	0

Fig. 1 Relations $\mathcal{R}$ and $\mathcal{R}^\bullet$

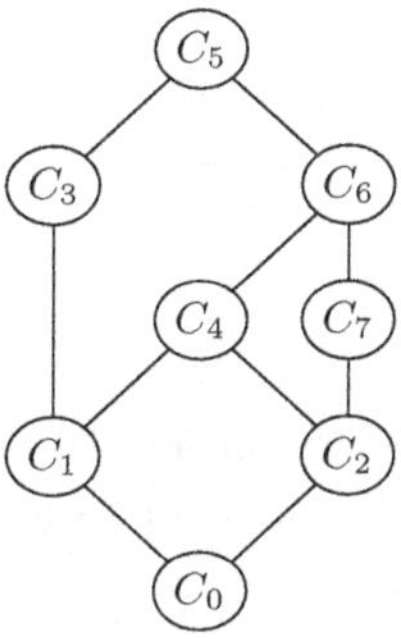

Concept	LS-extent	LS-intent
C_0	$\{\}$	$\{a, b, c, d, e, f\}$
C_1	$\{x_2\}$	$\{0.6/a, 0.7/b, c, 0.5/d, e, 0.7/f\}$
C_2	$\{x_3\}$	$\{a, 0.9/b, c, 0.1/d, e, f\}$
C_3	$\{x_1, x_2\}$	$\{0.5/a, 0.7/b, 0.7/c, 0.5/d, 0.7/e, 0.7/f\}$
C_4	$\{x_2, x_3\}$	$\{0.6/a, 0.7/b, c, 0.1/d, e, 0.7/f\}$
C_5	$\{x_1, x_2, x_3, x_4\}$	$\{0.5/a, 0.7/b, 0.7/c, 0.1/d, 0.7/e, 0.7/f\}$
C_6	$\{x_2, x_3, x_4\}$	$\{0.6/a, 0.7/b, 0.9/c, 0.1/d, 0.9/e, 0.7/f\}$
C_7	$\{x_3, x_4\}$	$\{a, 0.9/b, 0.9/c, 0.1/d, 0.9/e, 0.7/f\}$

Fig. 2 Concept lattice of the context in Example 1

Afterward, the scaled attributes are replaced by the original ones in the discernibility function. Furthermore, we will compute the reduced DNF:

$$\begin{aligned}(a \vee c \vee e) \wedge (a \vee b) \wedge (c \vee e) \wedge d &= (a \vee b) \wedge (c \vee e) \wedge d \\ &= (a \wedge c \wedge d) \vee (a \wedge d \wedge e) \vee (b \wedge c \wedge d) \vee (b \wedge d \wedge e)\end{aligned}$$

On the other hand, we will use the reduction mechanism of the multi-adjoint concept lattice framework in the original formal context. Therefore, we consider the multi-adjoint formal context $(\mathcal{O}, \mathcal{P}, \mathcal{R}, \sigma)$, where σ assigns the left-sided adjoint triple [1] on the granular interval $[0, 1]_{10}$ over the set of attributes.

The concept lattice associated with this context is shown in Fig. 2.

If we compute the concept lattice associated with the CR-formal context described in Table 1, we obtain the same concept lattice as the one shown in Fig. 2. As we can see in Table 1, the extents are the same as the concepts obtained in the left-sided concept lattice, as Theorem 1 asserted.

Nevertheless, the intents are not expressed in the same way, they are representing the same information. For example, we consider the label $\{a_{\geq 0.6}, c_{\geq 0.9}, e_{\geq 0.9}\}$, which belongs to the intent of concepts C_0, C_1, C_2, C_4, C_6 and C_7, we have that $X_l^{\uparrow}(a) \geq 0.6$, $X_l^{\uparrow}(c) \geq 0.9$ and $X_l^{\uparrow}(e) \geq 0.9$, according to Proposition 1.

Now, in order to illustrate Proposition 2, we will use the subset of attributes $\{x_2, x_3\}$. We have that $\{x_2, x_3\}^{\uparrow} = \{\{a_{\geq 0.6}, c_{\geq 0.9}, e_{\geq 0.9}\}, \{c_{\geq 1.0}, e_{\geq 1.0}\}\}$. As we can see in Fig. 2, we have that $X_l^{\uparrow}(c) = \max\{0.9, 1.0\} = 1.0$.

Table 1 Concepts of the CR-concept lattice

Concept	CR-extent	CR-intent
C_0	$\{\}$	$\{\{a_{\geq 0.6}, c_{\geq 0.9}, e_{\geq 0.9}\}, \{a_{\geq 1.0}, b_{\geq 0.9}\}, \{c_{\geq 1.0}, e_{\geq 1.0}\}, \{d_{\geq 0.5}\}\}$
C_1	$\{x_2\}$	$\{a_{\geq 0.6}, c_{\geq 0.9}, e_{\geq 0.9}\}, \{c_{\geq 1.0}, e_{\geq 1.0}\}, \{d_{\geq 0.5}\}\}$
C_2	$\{x_3\}$	$\{a_{\geq 0.6}, c_{\geq 0.9}, e_{\geq 0.9}\}, \{a_{\geq 1.0}, b_{\geq 0.9}\}, \{c_{\geq 1.0}, e_{\geq 1.0}\}\}$
C_3	$\{x_1, x_2\}$	$\{\{d_{\geq 0.5}\}\}$
C_4	$\{x_2, x_3\}$	$\{a_{\geq 0.6}, c_{\geq 0.9}, e_{\geq 0.9}\}, \{c_{\geq 1.0}, e_{\geq 1.0}\}\}$
C_5	$\{x_1, x_2, x_3, x_4\}$	$\{\}$
C_6	$\{x_2, x_3, x_4\}$	$\{\{a_{\geq 0.6}, c_{\geq 0.9}, e_{\geq 0.9}\}\}$
C_7	$\{x_3, x_4\}$	$\{\{a_{\geq 0.6}, c_{\geq 0.9}, e_{\geq 0.9}\}, \{a_{\geq 1.0}, b_{\geq 0.9}\}\}$

To conclude, the reducts obtained with the left-sided mechanism are $D_1 = \{a, c, d\}$, $D_2 = \{a, d, e\}$, $D_3 = \{b, c, d\}$ and $D_4 = \{b, d, e\}$. The same reducts are computed with the CR-method.

4 Conclusion and Future Work

Through this paper, the relation between the reduction mechanism in multi-adjoint framework and the CR-method has been studied. The extents of the concepts computed are equal in both theories. Specifically, a relation between the scaled values of the attributes in the CR-formal context and the values taken by the attributes in the left-sided framework has been presented. Furthermore, the obtained reducts are the same using both mechanisms. As a consequence, the multi-adjoint concept lattice reduction mechanism generalizes the CR-method in more flexible frameworks.

In the future, we will use this relation in order to connect the reduction mechanism in Rough Set Theory when the FCA mechanism is considered. Moreover, we will study how the reduction is affected when a specific value of the attribute is erased, instead of the whole attribute.

References

1. Benítez-Caballero, M.J., Medina, J., Ramírez-Poussa, E.: Characterizing one-sided formal concept analysis by multi-adjoint concept lattices. Mathematics **10**(7) (2022)
2. Cornejo, M.E., Medina, J., Ramírez-Poussa, E.: Characterizing reducts in multi-adjoint concept lattices. Inf. Sci. **422**, 364–376 (2018)
3. Cornejo, M.E., Medina, J., Ramírez-Poussa, E.: Algebraic structure and characterization of adjoint triples. Fuzzy Sets Syst. **425**, 117–139 (2021)
4. Ganter, B., Wille, R.: Formal Concept Analysis: Mathematical Foundation. Springer, Berlin (1999)

5. Konecny, J.: On attribute reduction in concept lattices: methods based on discernibility matrix are outperformed by basic clarification and reduction. Inf. Sci. **415**, 199–212 (2017)
6. Medina, J., Ojeda-Aciego, M., Ruiz-Calviño, J.: Formal concept analysis via multi-adjoint concept lattices. Fuzzy Sets Syst. **160**(2), 130–144 (2009)
7. Ren, R., Wei, L.: The attribute reductions of three-way concept lattices. Know.-Based Syst. **99**(C). 92–102 (2016)
8. Shao, M.W., Li, K.W.: Attribute reduction in generalized one-sided formal contexts. Inf. Sci. **378**, 317–327 (2017)
9. Shao, M.W., Yang, H.Z., Wu, W.Z.: Knowledge reduction in formal fuzzy contexts. Knowl.-Based Syst. **73**, 265–275 (2015)

ITAMACDSS: A Fuzzy Clinical Diagnostic Support System for Coeliac Disease

Marco Elio Tabacchi

1 Introduction

Coeliac disease (CD) is a genetically-induced disease that affects the human intestine, and is caused in predisposed subjects by frequent ingestion of a protein complex called gluten, that can be usually found in a number of commonly eaten cereals such as grain [1], to which byproducts it confers elasticity. CD is generally underdiagnosed in the general population, and is estimated that more than 1out of 100 adults suffers from the condition, up to in 1in 150 from 20 years ago. The CD diagnosis is a multi-step process, well codified in a series of medical protocols such as ESPGHAN (for children), and while a number of clues can be determined through these steps, the only certain way of diagnosis is biopsy of the intestine, an invasive procedure that has a significative cost both in terms of resources and of patient's discomfort, especially when it is carried out on children.

Project ITAMA, a collaboration between Italy and Malta, has developed a system for earlier diagnosis of CD in children, consisting in the combination of a Database and a Clinical Diagnostic Support System (CDSS) powered by Fuzzy Logic, fed by an data collection with more than 20.000 children as subjects. This paper presents a description of ITAMACDSS [2], as well as preliminary results obtained using data from the ITAMA project [3].

M. E. Tabacchi (✉)
DMI, Università Degli Studi Di Palermo, Palermo, Italy
e-mail: marcoelio.tabacchi@unipa.it

Istituto Nazionale Di Ricerche Demopolis, Palermo, Italy

M. E. Cornejo et al. (eds.), *Computational Intelligence and Mathematics for Tackling Complex Problems 5*, Studies in Computational Intelligence 1127,
https://doi.org/10.1007/978-3-031-46979-4_12

2 Description and Method

The classifier at the heart of ITAMACDSS is based on a fuzzy paradigm. The main advantages of using such a fremwork areas follows:

- as a definitive CD diagnosis is only possible through biopsy, the fact that the system generates approximate results a suggestions, and not imperative indications, should be of help for the clinicians to add its suggestions into the diagnostic pathways indicated in the protocols;
- using fuzzified values both as input as well as output in the system favours the introspective and explicable characteristics of the algorithm;
- the use of syntactic patterns more adapted to the semantic nature of the task allows for different explicata [4, 5];
- missing data, always present when such a vast data collection is performed due to abandonment of the experiment as well as other cuses, are more naturally represented in the Fuzzy context. The more the algorithm advances trough the steps of the pathway, the more the confidence about missing data will increase.

Such paradigm fits well with the idea that the diagnostic pathway for CD is an ongoing process that can follow many different paths, and not a batch job, that is done in its entirety before obtaining an answer. A classical binary classifier is well suited when it is possible to gather an (almost) complete set of data, such as blood tests, answers from anamnesis, other medical data, and then try to classify the patients to whom such data belongs as positive or negative. In CD, the diagnostic pathway [6] can be modified, and at each step is important to have at least a suggestion on how to better proceed. A fuzzy classifier can better represent this inherent uncertainty, by classifying a subject at each step of the pathway with different, usually increasing degrees of confidence as soon as more data is produced.

2.1 Implementation

The fuzziness variable based CDSS is built on a heavily tweaked version of a classic, feed-forward backpropagation with momentum neural network written by the author in the Python computer language, according to some of the ideas put forward in [7]. It differs from typical non-fuzzy binary classification as follows:

- input variables are encoded both as classical values, missing (when value is not known or cannot be measured) or fuzzy values, using any of a different number representations (triangular, trapezoidal, etc.), according to the appropriateness of the representation to the variable;
- fuzzy neurons are implemented using variations on the Model III idea in [7];
- adaptation of the weights of the neurons is carried out by weighting ('synaptic' adaptation), with limits pertaining to shifting and overlapping inspired from [8];

- a five-class response input from MIN to MAX, with maximal value MIN assigned to a confirmed negative subject and maximal MAX to a positive. Each output is constituted by an optimized fuzzy partition, using a five-classes response, determined by training and testing of the NN. When optimization is complete, each successive sample is run through the NN and assigned an output corresponding to the class to which the sample is determined to better belong;
- in accounting for the results, a true Positive is counted when a MAX is classified as a MAX, and a false positive when a MIN is classified as a MAX. The same is true when such variables are inverted, this time as false Positive and Negative. Other classifications are discarded, as not enough information is gained for a strong diagnosis suggestion.

2.2 Data Cleaning and Feature Extraction

The methodology described in [9] forms the base of the feature extraction phase of the algorithm. The input comes from ITAMA database, as collected by clinicians, and beforehand is cleaned from spurious data and verified for correctness:

- a 'zero model' using null variables and standard Fuzzy set is implemented;
- the training phase of the model is carried out on a subsets of the original data, randomly extracted, and the arithmetical mean of the output from such repetitions is taken as the result;
- each variable is tested separately, and the variable with the best accuracy computed in the previous phase is added to the list of variables for the final algorithm;
- the addition of variables is stopped when there is no more significant improvement in the classification, according to a certain threshold value gained by direct experimentation.
- once variable selections is competed, a gradient with momentums descent is applied to compute how many neurons each of the internal levels have.

The aforementioned algorithm selected the following parameters: 7 input neurons, two hidden layers with 5 and 3 neurons, one output layer; 200 epochs, 0.5 rate, 0.5 momentum, 0.05 epsilon. The chosen variables are: number of positive answers to the questions in the questionnaire, presence of chronic diarrhoea, presence of irregular bowel habits, results from the POC test, results from AAC test, total IGA, total IGG.

2.3 Risk and Confidence Factor

After the training phase, risk factor and confidence factor are computed for each new subject inserted into the DB.

The Risk Factor is the result that best fits the gathered information, while the Confidence Factor is a parameter that increases through time, taking into account the fact that running the diagnostic pathway adds information through the way. In prder to compute CF each stage of the pathway has a specific parameter w_i, derived by clinicians and medical literature. Each parameter is multiplied by a factor depending on the confidence of classification. The parameters in the first runs are as follows: w_1: 0.30, w_2: 0.55, w_3: 0.70. The update value is $w_i \cdot 2(fs_i)$. . All results are then normalised in [0, 1].

In the ITAMA system interface, RF and CF are converted into eidetic information that the DB system can display next to a patient's name in a patient list. This should be interpreted as a suggestion for the clinicians and health workers using the system. The RF is represented by colours (red, orange, yellow, yellow-green, green), the CF by a different colour transparency component. This information can be used by clinicians to prioritise patients who appear to be at higher risk for a positive diagnosis and to implement cost-saving measures during the screening process.

3 Results and Discussion

3.1 *Results*

A subset of patients enrolled in the ITAMA project screening campaign was used as a testbed for the classifier, omitting the ones with zero or missing data for al columns. The parameters used for training and testing are the ones described in the previous section. Training phase is done with a Pareto split (training-test). Due to the fact that there are much more negative than positive subjects—which itself is not a due to lack of data, but somehow reflects the ~1% prevalence of CD in the general European population—the ratio between positive and nepositive/negative ratio has been fixed in gative patients is kept at the same level of the general population with approximately 2% of positive cases. Prevalence of positive subjects in real data is higher than the accepted one for general European population, which may be due to different epidemiological factors such as geographical provenience, age and increasing incidence in recent times, subjects of further medical research.

Results in Table 1 are average from 1 K split batches (Monte Carlo repeating subsampling validation). True Positive mean: 23.78, standard deviation: 1.41; False Positive mean: 1.00, standard deviation: 0.17; False Negative mean: 3.67, standard deviation: 1.36; True Negative: 1399.76, standard deviation: 0.65). Performance values are: accuracy: 99%, sensitivity: 86%, specificity: 99%, precision: 96%. Computation time: O(n).

Table 1 Confusion matrix of averaged results from the ITAMACDSS classifier

	True (GT)	False (GT)
Predicted true	29	2
Predicted false	1	1583

3.2 Discussion

Results obtained by the classifier that underlies ITAMACDSS are comparable with the ones obtained by similar state-of-the-art research projects—for more on this see the review article [10], as well as [11–13]. The high values obtained allows to build a CDSS that can confidently help clinician assess the probability of being affected by CD in a patient and to act accordingly in the search for a correct diagnosis. More testing work remains to be done for a precise measuring of how well the RF and CF fit a posteriori with subjects' clinic profile, and how this information can be used to produce a cost-effective and patient-centred diagnostic pathway. Also due is a more detailed analysis of the neural network limits and characteristics, especially in borderline situations or with more complex input features.

Acknowledgements The present document is part of the ITAMA project [14] final documentation, prepared under the IC program INTERREG V-A (JS-Code C1-1.1-18). The content does not reflect the official opinion of the European Union, and responsibility for what is expressed therein expressed is entirely of the author.

References

1. Troncone, R., Jabri, B.: Coeliac disease and gluten sensitivity. J. Int. Med. **269**, 582–590 (2011)
2. Sorce, S., Giuliano, A., Tegolo, D., Gentile, V., Tabacchi, M. E., Valenti, C., Cascio, D., Taormina, V., Raso, G.: A REST-based framework to support non-invasive and early coeliac disease diagnosis. In: CompSysTech '19 Proceedings of the 20th International Conference on Computer Systems and Technologies, pp. 207–212 (2019)
3. Magazzù, G., Aquilina, S., Barbara, C., Bondin, R., Brusca, I., Bugeja, J., Camilleri, M., Cascio, D., Costa, S., Cuzzupè, C., Duca, A., Fregapane, M., Gentile, V., Giuliano, A., Grifò, A., Grima, A.M., Ieni, A., Li Calzi, G., Maisano, F., Melita, G., Pallio, S., Panasiti, I., Pellegrino, S., Romano, C., Sorce, S., Tabacchi, M.E., Taormina, V., Tegolo, D., Tortora, A., Valenti, C., Vella, C., Raso, G.: Recognizing the emergent and submerged iceberg of the celiac disease: ITAMA project-global strategy protocol. Pediatric Rep. **14**(2), 293–311 (2022)
4. Ahmadi, H., Gholamzadeh, M., Shahmoradi, L., Nilashi, M.: Pooria Rashvand, "Diseases diagnosis using fuzzy logic methods: a systematic and meta-analysis review." Comput. Methods Programs Biomed.. Methods Programs Biomed. **161**, 145–172 (2018). https://doi.org/10.1016/j.cmpb.2018.04.013
5. Minutolo, A., Esposito, M., De Pietro, G.: A fuzzy framework for encoding uncertainty in clinical decision-making. Knowl. Based Syst. **98**, 95–116 (2016)
6. Husby, S., Koletzko, S., Korponay-Szabo, I.R., Mearin, M.L., Phillips, A., Shamir, R., et al.: European society for pediatric gastroenterology, hepatology, and nutrition guidelines for the diagnosis of coeliac disease. J. Pediatr. Gastroenterol. Nutr.Pediatr. Gastroenterol. Nutr. **54**(1), 136–160 (2012). https://doi.org/10.1097/MPG.0b013e31821a23d0
7. Gupta, M.M.: The fuzzy neural network: the emerging paradigms. IFAC Proc. **25**(6), 127–129 (1992)
8. Kwan, H.K., Cai, Y.: A fuzzy neural network and its application to pattern recognition. IEEE Trans. Fuzzy Syst. **2**(3), 185–193 (1994). https://doi.org/10.1109/91.298447
9. Pota, M., Esposito, M., De Pietro, G.: Designing rule-based fuzzy systems for classification in medicine. Knowl.-Based Syst. **124**, 105–132 (2017)

10. Molder, A., Vasile Balaban, D., Jinga, M., Molder, C.C.: Current evidence on computer-aided diagnosis of celiac disease: systematic review. Front. Pharm. (2020)
11. Wanga, X., Qiana, H., Ciaccio, E.J., et al.: Celiac disease diagnosis from videocapsule endoscopy images with residual learning and deep feature extraction. Comput. Methods Programs Biomed. **187** (2020). https://doi.org/10.1016/j.cmpb.2019.105236.
12. Amirkhani, A., Mosavi, M.R., Mohammadi, K., Papageorgiou, E.I.: A novel hybrid method based on fuzzy cognitive maps and fuzzy clustering algorithms for grading celiac disease. Neural Comput. Appli **30**, 1573–1588 (2018)
13. Thukral, S., Singh Bal, J.: Fuzzy logic: an easiest technique to predict celiac disease. Sci. Technol. J. **7**(2) (2019)
14. Tabacchi, M.E., et al.: A fuzzy-based clinical decision support system for coeliac disease. IEEE Access **10**, 102223–102236 (2022). https://doi.org/10.1109/ACCESS.2022.3208903

Preference in Multi-adjoint Logic Programming Based on Ordered Adjoint Pairs

Jesús Medina and José Antonio Torné-Zambrano

1 Introduction

Preferences in decision-making have formally been treated from a mathematical point of view since the arrival of the notion of fuzzy set [11]. Thanks to membership functions (and taking into account the framework of decision-making problems based on the presence of uncertainty and lack of information), fuzzy sets have shown to be an adequate way to formalize the abstract notion of *preference*, as studied by Dubois and Prade in their papers about the three semantics of fuzzy sets [3–5]. All these papers based on decision making analysis has been studied together under the name of Fuzzy Decision Analysis (FDA).

The fuzzy logic programming framework considered in this paper will be the multi-adjoint one [1, 8], which models more complex situations than other logic programming frameworks. One of the main advantages of multi-adjoint logic programming is the possibility of taking into account diverse adjoint pairs in the design of a fuzzy logical program.

This paper complements the preliminary work given in [2] with the primary objective of analyzing the impact of the use of different implications in the rules of a multi-adjoint logic program, in order to study preferences in decision-making frameworks.

Partially supported by the 2014–2020 ERDF Operational Programme in collaboration with the State Research Agency (AEI) in projects PID2019-108991GB-I00 and PID2022-137620NB-I00, and with the Department of Economy, Knowledge, Business and University of the Regional Government of Andalusia in project FEDER-UCA18-108612, and by the European Cooperation in Science & Technology (COST) Action CA17124.

J. Medina · J. A. Torné-Zambrano (✉)
Department of Mathematics, University of Cádiz, Cádiz, Spain
e-mail: joseantonio.torne@uca.es

J. Medina
e-mail: jesus.medina@uca.es

M. E. Cornejo et al. (eds.), *Computational Intelligence and Mathematics for Tackling Complex Problems 5*, Studies in Computational Intelligence 1127,
https://doi.org/10.1007/978-3-031-46979-4_13

A study of how this choice affects the computation of the minimum models will be carried out, including different results. Moreover, some examples to illustrate the use of different (ordered) adjoint pairs will also be introduced.

2 Preliminaries

First of all, we recall some definitions useful in the understanding of the following sections. For more details see [8, 9].

Definition 1 Let $(P, \leq_P)$ and $(Q, \leq_Q)$ be two posets, A a set and the operators $\&\colon P \times A \to Q$ and $\swarrow\colon Q \times A \to P$. We say that $(\&, \swarrow)$ is a *right adjoint pair* w.r.t. P, A, Q if the following condition is verified

$$x \leq_P z \swarrow y \quad \text{iff} \quad x \& y \leq_Q z$$

for all $x \in P$, $y \in A$ and $z \in Q$. This pair will be simply called *adjoint pair* in the multi-adjoint logic programming framework.

The following example shows the main adjoint pairs we are going to consider in this study.

Example 1 The adjoint pairs $(\&_G, \leftarrow_G)$, $(\&_Ł, \leftarrow_Ł)$ and $(\&_p, \leftarrow_p)$, where $\&_G$, $\&_Ł$ and $\&_p$ are the Gödel, Łukasiewicz and product adjoint conjunctor respectively, are defined, for all $x, y, z \in [0, 1]$, as

$$\&_G(x, y) = \min\{x, y\} \qquad z \leftarrow_G y = \begin{cases} 1 & \text{if } y \leq z \\ z & \text{otherwise} \end{cases}$$

$$\&_Ł(x, y) = \max\{0, x + y - 1\} \qquad z \leftarrow_Ł y = \min\{1, 1 - y + z\}$$

$$\&_p(x, y) = x \cdot y \qquad z \leftarrow_p y = \begin{cases} 1 & \text{if } y \leq z \\ \frac{z}{y} & \text{otherwise} \end{cases}$$

□

Finally, the definition of the immediate consequence operator is recalled, which is given on a definite and propositional fuzzy logic program [8].

Definition 2 Let $\mathbb{P}$ be a multi-adjoint logic program. The *immediate consequence operator* $T_{\mathbb{P}}^{\mathfrak{L}}\colon \mathcal{I}_{\mathfrak{L}} \longrightarrow \mathcal{I}_{\mathfrak{L}}$, mapping interpretations to interpretations, is defined by considering

$$T_{\mathbb{P}}^{\mathfrak{L}}(I)(A) = \sup\left\{\vartheta \dot{\&}_i \hat{I}(\mathcal{B}) \mid A \overset{\vartheta}{\leftarrow}_i \mathcal{B} \in \mathbb{P}\right\}$$

3 Studying the Impact of Different Implications in Multi-adjoint Logic Programming

From now on, we will use the notation $\mathbb{P}_{ni}$ or $\mathbb{P}_{n\&_i}$ for a logic program in which every implication used in the construction of the n rules is the same, that is, $\leftarrow_i$, with their respective adjoint conjunctor $\&_i$. For example, $\mathbb{P}_{nG}$ is a multi-adjoint logic program with n rules, in which every rule is formed using the Gödel implication $\leftarrow_G$. Analogously, we could define logic programs $\mathbb{P}_{np}$ and $\mathbb{P}_{nŁ}$, with the product and Łukasiewicz conjunctors. We will use the notation $\mathbb{P}_{nij}$ or $\mathbb{P}_{n\&_i\&_j}$ for an arbitrary logic program considering the conjunctors $\&_i$ and $\&_j$. We will consider in this paper the linear complete lattice $([0,1], \leq)$.

The first result is focused on the 'monotonicity' of the immediate consequence operator with respect to the considered adjoint pairs in the rules of the multi-adjoint program. Specifically, it shows that greater adjoint conjunctors provide greater immediate consequence operators.

Theorem 1 *Let* $\&_1\colon [0,1] \times [0,1] \to [0,1]$ *and* $\&_2\colon [0,1] \times [0,1] \to [0,1]$ *be two adjoint conjunctors verifying that* $\&_1 \leq \&_2$, *that is,* $x \&_1 y \leq x \&_2 y$, *for all* $(x, y) \in [0,1] \times [0,1]$. *Then, it is satisfied that* $T_{\mathbb{P}_{n\&_1}} \leq T_{\mathbb{P}_{n\&_2}}$.

As a consequence, we obtain the following chain for the usual Gödel, Łukasiewicz and product t-norms.

Corollary 1 *It is verified that* $T_{\mathbb{P}_{nŁ}} \leq T_{\mathbb{P}_{nP}} \leq T_{\mathbb{P}_{nG}}$.

In the following example we are going to study the effect on the value of the immediate consequence operator changing the implication used on the rule.

Example 2 Consider the program $\mathbb{P}_{2GŁ}$ formed with the rules $\langle A \leftarrow_G B_1, \vartheta_1\rangle$ and $\langle A \leftarrow_Ł B_2, \vartheta_2\rangle$ and the program $\mathbb{P}_{2pŁ}$ composed of the rules $\langle A \leftarrow_p B_1, \vartheta_1\rangle$ and $\langle A \leftarrow_Ł B_2, \vartheta_2\rangle$. The immediate consequence operator can be obtained, in both cases, as: $T_{\mathbb{P}_{2GŁ}}(I)(A) = \sup\left\{\vartheta_1 \dot{\&}_G \hat{I}(B_1), \vartheta_2 \dot{\&}_Ł \hat{I}(B_2)\right\}$ and $T_{\mathbb{P}_{2pŁ}}(I)(A) = \sup\left\{\vartheta_1 \dot{\&}_p \hat{I}(B_1), \vartheta_2 \dot{\&}_Ł \hat{I}(B_2)\right\}$. Therefore, the impact of the rule (relation among B_1 and A) in the modeled system is downplayed. However, depending on the interpretation I, it could not be effective as we show next. We can establish the following casuistry.

- If $\vartheta_2 \dot{\&}_Ł \hat{I}(B_2) \leq \vartheta_1 \dot{\&}_G \hat{I}(B_1)$, then it is verified that $\vartheta_1 \dot{\&}_p \hat{I}(B_1) \leq \vartheta_1 \dot{\&}_G \hat{I}(B_1)$ and, consequently $T_{\mathbb{P}_{2pŁ}} \leq T_{\mathbb{P}_{2GŁ}}$. Therefore, if a Gödel implication is replaced by a product one, then the change of the implication has a real impact in the computations obtaining lower values.
- If $\vartheta_1 \dot{\&}_G \hat{I}(B_1) \leq \vartheta_2 \dot{\&}_Ł \hat{I}(B_2)$, then $\vartheta_1 \dot{\&}_p \hat{I}(B_1) \leq \vartheta_2 \dot{\&}_Ł \hat{I}(B_2)$ and $T_{\mathbb{P}_{2GŁ}} = T_{\mathbb{P}_{2pŁ}}$. Consequently, the change of implication does not affect the value of the immediate consequence operator.
- If $\vartheta_2 \dot{\&}_Ł \hat{I}(B_2) \leq \vartheta_1 \dot{\&}_p \hat{I}(B_1)$, then it is verified that $\vartheta_1 \dot{\&}_Ł \hat{I}(B_1) \leq \vartheta_1 \dot{\&}_G \hat{I}(B_1)$ and $T_{\mathbb{P}_{2pŁ}} \leq T_{\mathbb{P}_{2GŁ}}$. The value of the immediate consequence operator would change

from $\vartheta_1 \dot{\&}_p \hat{I}(B_1)$ to $\vartheta_1 \dot{\&}_G \hat{I}(B_1)$. Therefore, the importance of the rule $\langle A \leftarrow B_1, \vartheta_1 \rangle$ is obvious.
- If $\vartheta_1 \dot{\&}_p \hat{I}(B_1) \leq \vartheta_1 \dot{\&}_Ł \hat{I}(B_2)$, then nothing can be deduced from this case. □

Theorem 2 *Consider two adjoint conjunctors* $\&_1 \colon [0, 1] \times [0, 1] \to [0, 1]$ *and* $\&_2 \colon [0, 1] \times [0, 1] \to [0, 1]$ *such that* $\&_1 \leq \&_2$. *Then, it is verified that* $T_{\mathbb{P}_{n\&_1}} \leq T_{\mathbb{P}_{n\&_1\&_2}} \leq T_{\mathbb{P}_{n\&_2}}$.

The following corollary shows some chains considering usual t-norms.

Corollary 2 *Consider a multi-adjoint logic program* $\mathbb{P}$ *and the family of adjoint conjunctors* $\Lambda = \{\&_Ł, \&_p, \&_G\}$, *together with their respective residuated implications. Then, it is verified that*

1. $T_{\mathbb{P}_{n\&_Ł}} \leq T_{\mathbb{P}_n} \leq T_{\mathbb{P}_{n\&_G}}$, *for all* $n \in \mathbb{N}$, *where* $T_{\mathbb{P}_n}$ *denotes a program with any possible combination of residuated implications of elements in* Λ.
2. $T_{\mathbb{P}_{n\&_p\&_Ł}} \leq T_{\mathbb{P}_{n\&_p}} \leq T_{\mathbb{P}_{n\&_p\&_G}}$.

Corollary 3 *Consider a multi-adjoint logic program* $\mathbb{P}$ *and the family of conjunctors* $T = \{\&_1, \&_2, \ldots, \&_t\}$, *with* $t \in \mathbb{N}, t \geq 2$, *verifying that* $\&_1 \leq \&_2 \leq \ldots \leq \&_t$. *Then, it is satisfied that*

$$T_{\mathbb{P}_{n\&_1}} \leq T_{\mathbb{P}_{n\&_1\&_2}} \leq T_{\mathbb{P}_{n\&_2}} \leq \ldots \leq T_{\mathbb{P}_{n\&_{t-1}}} \leq T_{\mathbb{P}_{n\&_{t-1}\&_t}} \leq T_{\mathbb{P}_{n\&_t}}$$

In the following example, we will apply these results to a particular case in which we use a logic program in order to distinguish between covid and cold.

Example 3 Consider the multi-adjoint logic program $\mathbb{P}$ composed of the following rules.

covid $\overset{0.5}{\leftarrow}$ diarrhoea $\wedge_Ł$ headache
covid $\overset{0.9}{\leftarrow}$ moderate fever
covid $\overset{0.8}{\leftarrow}$ loss of smell and taste
covid $\overset{0.4}{\leftarrow}$ sore throat $\wedge_Ł$ muscle pain
covid $\overset{0.7}{\leftarrow}$ cough
covid $\overset{0.6}{\leftarrow}$ stuffy and runny nose $\wedge_Ł$ shortness of breath
covid $\overset{0.6}{\leftarrow}$ fatigue

cold $\overset{0.6}{\leftarrow}$ diarrhoea $\wedge_Ł$ moderate fever
cold $\overset{0.1}{\leftarrow}$ shortness of breath
cold $\overset{0.8}{\leftarrow}$ fatigue $\wedge_Ł$ muscle pain
cold $\overset{0.5}{\leftarrow}$ loss of smell and taste $\wedge_Ł$ sore throat
cold $\overset{0.6}{\leftarrow}$ stuffy and runny nose
cold $\overset{0.9}{\leftarrow}$ cough
cold $\overset{0.7}{\leftarrow}$ headache

These rules are completed with observations about each particular person, in the form of *facts*. For example, one specific individual can show the following symptoms.

moderate fever $\overset{0.6}{\leftarrow} \top$
shortness of breath $\overset{0.1}{\leftarrow} \top$
loss of smell and taste $\overset{0.1}{\leftarrow} \top$
stuffy and runny nose $\overset{0.7}{\leftarrow} \top$

sore throat $\overset{0.5}{\leftarrow} \top$
cough $\overset{0.8}{\leftarrow} \top$
fatigue $\overset{0.4}{\leftarrow} \top$

diarrhoea $\overset{0.6}{\leftarrow} \top$
headache $\overset{0.7}{\leftarrow} \top$
muscle pain $\overset{0.2}{\leftarrow} \top$

Taking into account the current health crisis and the great ease with which a person can get covid-19, it is advisable to do the diagnosis in secure areas, without risking of going to a clinic. With this purpose, multi-adjoint logic programming can be used in order to suggest a first diagnosis of the possible disease of the user. In this case, the introduced program will help to identify the possible disease of the patient, distinguishing between covid-19 and cold. We must bear in mind that this is a simple example with which to illustrate in the most didactic way the use of preferences in multi-adjoint logic programming. By Corollary 2, we can associate different degrees of preference with the conjunctors $\&_G$, $\&_Ł$ and $\&_P$.

- $\&_Ł$ is a low degree of preference. In a decision-making problem, $\&_Ł$ will be associated with the idea of "*I prefer not*".
- $\&_P$ is a medium degree of preference, corresponding with the idea of "*I neither prefer it nor stop preferring it*".
- $\&_G$ is a high degree of preference. In a decision-making problem, $\&_G$ will be associated with the idea of "*I prefer*".

Now, we are going to understand this problem in different contexts.

• First situation. We can assume that the user shows the symptoms aforementioned, but this person has barely left his house in the last two weeks, because he works on-line. He has only gone for shopping and walking. In addition, he met with a friend one week ago and his wife, who is a teacher, strictly complies with all security measures against covid-19. Therefore, at first, it is more probable that this person has a cold than covid-19, but we cannot discard this last possibility. Consequently, taking into account that his "degree of preference" of cold is higher, then in this situation we are going to use Gödel conjunctors for the rules with head `cold`, while Łukasiewicz one for the rules with head `covid`. Next, we calculate the value of the immediate consequence operator for the propositional symbol `covid`. We recall how the least fixed point of the immediate consequence operator is calculated starting from the least interpretation. The first is trivially: $T_{\mathbb{P}_Ł}(\triangle)(\texttt{covid}) = 0$.

$$
\begin{aligned}
T^2_{\mathbb{P}_Ł}(\triangle)(\texttt{covid}) &= T_{\mathbb{P}_Ł}\left(T_{\mathbb{P}_Ł}(\triangle)(\texttt{covid})\right)\\
&= \sup\Big\{0.5\dot{\&}_Ł \widehat{T_{\mathbb{P}_Ł}(\triangle)}(\texttt{diarrhoea} \wedge_Ł \texttt{headache}),\\
&\quad 0.9\dot{\&}_Ł \widehat{T_{\mathbb{P}_Ł}(\triangle)}(\texttt{moderate fever}), 0.7\dot{\&}_Ł \widehat{T_{\mathbb{P}_Ł}(\triangle)}(\texttt{cough}),\\
&\quad 0.8\dot{\&}_Ł \widehat{T_{\mathbb{P}_Ł}(\triangle)}(\texttt{loss of smell and taste}),\\
&\quad 0.4\dot{\&}_Ł \widehat{T_{\mathbb{P}_Ł}(\triangle)}(\texttt{sore throat} \wedge_Ł \texttt{muscle pain}),\\
&\quad 0.6\dot{\&}_Ł \widehat{T_{\mathbb{P}_Ł}(\triangle)}(\texttt{stuffy and runny nose}\\
&\qquad \wedge_Ł \texttt{shortness of breath}),\\
&\quad 0.6\dot{\&}_Ł \widehat{T_{\mathbb{P}_Ł}(\triangle)}(\texttt{fatigue})\Big\}\\
&= \sup\left\{0.5\dot{\&}_Ł(0.6 \wedge_Ł 0.7), 0.9\dot{\&}_Ł 0.6, 0.7\dot{\&}_Ł 0.8, 0.4\dot{\&}_Ł(0.5 \wedge_Ł 0.2),\right.\\
&\quad \left. 0.8\dot{\&}_Ł 0.1, 0.6\dot{\&}_Ł(0.7 \wedge_Ł 0.1), 0.6\dot{\&}_Ł 0.4\right\}\\
&= \sup\{0, 0.5, 0.5, 0, 0, 0, 0\} = 0.5
\end{aligned}
$$

Analogously, it can be shown that $T_{\mathbb{P}^3_{Ł}}(\triangle)(\texttt{covid}) = 0.5$, $T_{\mathbb{P}_G}(\triangle)(\texttt{covid}) = 0$, $T_{\mathbb{P}^2_G}(\triangle)(\texttt{covid}) = 0.8$ and $T_{\mathbb{P}^3_G}(\triangle)(\texttt{covid}) = 0.8$. We can see that, in this situation, it is more probable that the person has cold than covid-19 because the value of the immediate consequence operator of `cold` (with the Gödel implication) is much higher than the value associated with `covid`.

• *Second situation*. We are going to proceed in reverse of the previous case. We can assume that the person was in contact with another person, without using the masks and without keeping the safety distance. Given these circumstances, it could be that the person thought that he could have been infected because of his friend, so the "preference degree" is now going to be covid-19, although it cannot be ruled out that it could also be the cold. Therefore, in the rules with head `covid`, we will use Gödel implications, while in the rules with head `cold`, Łukasiewicz ones. In this case, we can check that the value obtained is the same. Therefore, although the possibility that this person may present coronavirus rises (considerably when compared to using the other conjunctors), it is not enough to overcome the possibility of cold.

We obtain that $T_{\mathbb{P}_G}(\triangle)(\texttt{covid}) = 0$, $T_{\mathbb{P}^2_G}(\triangle)(\texttt{covid}) = 0.7$, $T_{\mathbb{P}^3_G}(\triangle)(\texttt{covid}) = 0.7$, $T_{\mathbb{P}_{Ł}}(\triangle)(\texttt{cold}) = 0$, $T_{\mathbb{P}^2_{Ł}}(\triangle)(\texttt{cold}) = 0.7$ and $T_{\mathbb{P}^3_{Ł}}(\triangle)(\texttt{cold}) = 0.7$.

• *Third situation*. Finally, if we are not sure about one of the symptoms, for example, rules with `fatigue` and `muscle pain`, then we can use in those rules a Łukasiewicz conjunctor, while in the rest of the rules we can consider a product or Gödel one. □

4 Conclusions and Future Work

The impact of the use of different ordered adjoint pairs has been introduced in this paper. This preliminary study has shown this ordering is naturally translated into the computation of the minimal model. Furthermore, an illustrative example has been presented containing some ideas about how the selections of adjoint pairs, depending on the preference of the user, can be done. In the future, more properties and interpretations of the selections will be studied. For instance, the relationship with the theory introduced in [6, 7, 10] will be analyzed.

References

1. Cornejo, M.E., Medina, J., Ramírez-Poussa, E.: Multi-adjoint algebras versus non-commutative residuated structures. Int. J. Approx. Reason. **66**, 119–138 (2015)
2. Cornejo, M.E., Medina, J., Ramírez-Poussa, E., Rubio-Manzano, C.: Multi-adjoint concept lattices, preferences and Bousi prolog. In: Flores, V., Gomide, F., Janusz, A., Meneses, C., Miao, D., Peters, G., Ślęzak, D., Wang, G., Weber, R., Yao, Y. (eds.) Rough Sets, pp. 331–341. Springer International Publishing, Cham (2016)

3. Dubois, D.: The role of fuzzy sets in decision sciences: old techniques and new directions. Fuzzy Sets Syst. **184**(1), 3–28 (2011). Preference Modelling and Decision Analysis (Selected Papers from EUROFUSE 2009)
4. Dubois, D., Prade, H.: The three semantics of fuzzy sets. Fuzzy Sets Syst. **90**(2), 141–150 (1997). Fuzzy Sets: Where Do We Stand? Where Do We Go?
5. Dubois, D., Prade, H.: Gradualness, uncertainty and bipolarity: making sense of fuzzy sets. Fuzzy sets Syst. **192**, 3–24 (2012)
6. Eiter, T., Fink, M.: Uniform Equivalence of Logic Programs Under the Stable Model Semantics. vol. 2916, pp. 224–238 (2003)
7. Lifschitz, V., Pearce, D., Valverde, A.: Strongly equivalent logic programs. ACM Trans. Comput. Log. **2**, 526–541 (2001)
8. Medina, J., Ojeda-Aciego, M., Vojtáš, P.: Multi-adjoint logic programming with continuous semantics. In: Logic Programming and Non-monotonic Reasoning, LPNMR'01. Lecture Notes in Artificial Intelligence, vol. 2173, pp. 351–364 (2001)
9. Medina, J., Ojeda-Aciego, M., Vojtáš, P.: Similarity-based unification: a multi-adjoint approach. Fuzzy Sets Syst. **146**, 43–62 (2004)
10. Woltran, S.: Characterizations for Relativized Notions of Equivalence in Answer Set Programming, vol. 3229, pp. 161–173 (2004)
11. Zadeh, L.A.: Fuzzy sets. Inf. Control **8**, 338–353 (1965)

Lattice Valued Aggregation Operators Applicable in Digital Forensics

M. Eugenia Cornejo, Jesús Medina, Ivana Štajner-Papuga, and Andreja Tepavčević

1 Introduction

The problem of ranking objects with respect to pre-set conditions is one contemporary topic [11, 17]. The conditions in question model the perception and preferences of experts and may differ depending on what the focus of the investigation is. Motivation for the work presented in this paper can be found in the method for data ranking proposed in [5] where the measure of interaction between the segments of the investigated data has been introduced.

Partially supported by 2014–2020 ERDF Operational Programme in collaboration with the State Research Agency (AEI) in projects PID2019-108991GB-I00 and PID2022-137620NB-I00, and with the Department of Economy, Knowledge, Business and University of the Regional Government of Andalusia in project FEDER-UCA18-108612, by the Ministry of Science, Technological Development and Innovation of the Republic of Serbia (Grant No. 451-03-47/2023-01/200125) and by the European Cooperation in Science & Technology (COST) Action CA17124.

M. E. Cornejo · J. Medina
Department of Mathematics, University of Cádiz, Cadiz, Spain
e-mail: mariaeugenia.cornejo@uca.es

J. Medina
e-mail: jesus.medina@uca.es

I. Štajner-Papuga (✉) · A. Tepavčević
Department of Mathematics and Informatics, Faculty of Sciences, University of Novi Sad, Novi Sad, Serbia
e-mail: ivana.stajner-papuga@dmi.uns.ac.rs

A. Tepavčević
e-mail: andreja@dmi.uns.ac.rs

A. Tepavčević
Mathematical Institute SANU, Belgrade, Serbia

M. E. Cornejo et al. (eds.), *Computational Intelligence and Mathematics for Tackling Complex Problems 5*, Studies in Computational Intelligence 1127,
https://doi.org/10.1007/978-3-031-46979-4_14

The main idea of this paper is to team up two well-known mathematical notions, bounded lattices and aggregation operators, in order to obtain a new powerful mathematical tool applicable to different practical areas, particularly for obtaining the appropriate measure of interaction that will reflect the opinions of experts in digital forensics.

Aggregation operators can be used for the extraction of a representative value from some collection of data. In general, an aggregation operator is an n-ary monotone mapping on the unit interval that fulfills boundary conditions [2], and, recently, aggregation operators are being developed and intensively investigated within the framework of bounded lattices (e.g. [9, 14]).

On the other hand, lattices are well-known algebraic structures with ordering relations in their core. Real numbers together with the usual ordering provides one of the most used (linear) lattices, although what is of special interest for further research is that some elements can be incomparable. More precisely, lattices allow the comparison of various aspects of objects, e.g., one object might be more convenient in one feature and less convenient in another feature, while the situation for the second object is reversed, so these two objects would be incomparable. Usually, it is required to have a least element and a greatest element in the lattice, obtaining a bounded lattice. For example, the unit interval [0, 1] is a bounded lattice that is used a lot in non-classical logics.

In this study, given a family of binary lattice/valued aggregation operators on a bounded lattice, a special type of n-ary lattice-valued aggregation operator will be introduced, which complements the study introduced in [5]. In addition, this is a generalization and upgrade of the results presented in [6] where this framework is developed for aggregation operators on the unit interval. The obtained operators will be used as a core part of the method for ranking different crimes or data, which was originally proposed in [5]. The method in question consists of four steps that link fuzzy sets [18], fuzzy measures [1, 16], and integral aggregation operators [7, 8]. A fuzzy measure is used to describe the importance and interaction of segments of data that need to be evaluated and model the opinion of an investigator or an expert. While the importance, i.e., fuzzy measure, of singletons is predefined by an expert, the measure of interaction, i.e., a fuzzy measure of non-singletons, is being calculated. Since each additional segment needs to have a stronger effect on the interaction, lattice-valued aggregation operators that can reflect that property are the main topic of this paper. This lattice approach is of special interest for further research due to its adaptability in situations where some data is missing.

2 Preliminary Notions

The general definition of an aggregation operator, given through the framework of bounded lattices, is recalled next. As it can be seen from the definition itself, this general form is applicable to an arbitrary number of input values.

Definition 1 Given a bounded lattice L with the top and the bottom elements, denoted as 1 and 0, respectively, an *extended lattice-valued aggregation operator* is an extended mapping

$$A\colon \bigcup_{n\in\mathbb{N}} L^n \longrightarrow L$$

such that, $A(x) = x$, for all $x \in L$, and its restriction to L^n, is monotone in each variable, and fulfills boundary conditions $A(0, 0, \ldots, 0) = 0$, and $A(1, 1, \ldots, 1) = 1$, for all $n \in \mathbb{N}$.

In general, aggregation operators need not be commutative or associative.

3 Lattice Valued *n*-ary Aggregation Operator Associated with Binary Aggregation Operators

Based on the motivation and the idea presented in [5] for obtaining monotone n-ary aggregation operators from binary ones, this section introduced the formal definition of these n-ary aggregation operators on bounded lattices.

Definition 2 Let $n \geq 2$ and $A\colon L^n \longrightarrow L$ be an n-ary lattice mapping, monotone in all of the arguments, satisfying $A(x, 0 \ldots, 0) = x$, and

$$\bigvee_{i=1}^{n} x_i \leq A(x_1, \ldots, x_n)$$

Then, the *family of binary operators associated with* A, denoted as $\{@_i\colon L^2 \to L \mid i \in \{1, \ldots, n-1\}\}$, is of the following form, for all $x, y \in L$

$$\begin{aligned}
@_1(x, y) &= A(x, y, 0, \ldots, 0)\\
@_2(x, y) &= A(x, y, y, \ldots, 0)\\
&\vdots\\
@_{n-1}(x, y) &= A(x, y, y, \ldots, y)
\end{aligned}$$

It should be stressed that operators given by the previous definition are not commutative, which is actually a very convenient feature in applications that request that the importance of the order of the input segments should be emphasized.

Given $x_1, x_2, \ldots, x_n \in L$, and an n-ary lattice mapping A, from the condition $\bigvee_{i=1}^{n} x_i \leq A(x_1, \ldots, x_n)$, it follows that 1 is an absorptive element, i.e., $A(x_1, \ldots, x_n) = 1$ whenever at least one of the elements $x_1, \ldots, x_n$ is equal to 1. Similarly, 1 is absorptive element for each $@_i$. In particular, $A(1, 1, \ldots, 1) = 1$ and $@_i(1, 1) = 1$. Moreover, $A(0, 0 \ldots, 0) = 0$, and $@_i(0, 0) = 0$ are satisfied. Therefore, A and every $@_i$, with $i \in \{1, \ldots, n-1\}$, are aggregation operators in the sense of [7], i.e., monotonicity and boundary conditions hold.

Proposition 1 *If the family of mappings* $\{@_i \colon L^2 \to L \mid i \in \{1, \ldots, n-1\}\}$, *is given as in Definition 2, then for each* $i \in \{1, \ldots, n-1\}$ *we have that* 0 *is the right neutral element* ($@_i(x, 0) = x$, *for all* $x, y \in L$), $x \vee y \leq @_i(x, y)$, *it is monotone in both arguments for all* $x, y \in L$, *and the following holds*

$$@_1(x, y) \leq @_2(x, y) \leq \cdots \leq @_{n-1}(x, y).$$

An opposite proposition can also be formulated and proved.

Proposition 2 *Let* $\{@_i \colon L^2 \to L \mid i \in \{1, \ldots, n-1\}\}$ *be a family of monotone mappings satisfying that* $@_i(x, 0) = 0$ *and* $x \vee y \leq @_i(x, y)$, *for all* $i \in \{1, \ldots, n-1\}$, $x, y \in L$, *and the following chain of inequalities* $@_1(x, y) \leq @_2(x, y) \leq \cdots \leq @_{n-1}(x, y)$ *holds. Let* $A_@ \colon L^n \longrightarrow L$ *be a mapping defined as*

$$A_@(x_1, x_2, \ldots, x_n) = @_{n-1}(@_{n-2}(\ldots @_2(@_1(x_1, x_2), x_3), \ldots, x_{n-1}), x_n)$$

Then, $A_@$ *is monotone in all of the arguments and satisfies*

$$\bigvee_{i=1}^{n} x_i \leq A_@(x_1, \ldots, x_n)$$

and $A_@(x, 0, \ldots, 0) = x$, *for all* $x, x_1, x_2, \ldots, x_n \in L$.

The following proposition considers the special case of idempotent operators.

Proposition 3 *Given the mapping* $A \colon L^n \to L$, *defined as in Definition 2. If* A *is idempotent, i.e.,* $A(x, x, \ldots, x) = x$, *for all* $x \in L$, *then*

- *(i)* $A(x_1, \ldots, x_n) = \bigvee_{i=1}^{n} x_i$, *for all* $x_1, \ldots, x_n \in L$.
- *(ii) Each element of the family of binary operators associated with* A *is idempotent. Consequently,* $@_i(x, y) = x \vee y$, *for all* $i \in \{1, \ldots, n-1\}$ *and* $x, y \in L$.
- *(iii) The operator* $A_@$, *obtained from the family of binary operators associated with* A, *is identical to* A.

Thus, in the idempotent setting, the operator $A_@$ is identical to the starting operator A. Furthermore, we also have that, if the starting point is an idempotent family of binary operations satisfying the requirements of Proposition 2, then this family is equal to the family of binary operators associated with A. However, this does not happen in general.

Lemma 1 *Let* $\{@_i \colon L^2 \to L \mid i \in \{1, \ldots, n-1\}\}$ *be a family of binary operators satisfying conditions from Proposition 2, and let* $A_@$ *be the* n*-ary operator obtained in Proposition 2. Let* $\{\overline{@_i} \colon L^2 \to L \mid i \in \{1, \ldots, n-1\}\}$ *be the family of binary operators obtained by Definition 2 from* $A_@$. *Then, the following inequalities hold for all* $x, y \in L$:

$$@_1(x, y) = \overline{@_1}(x, y) \quad \text{and} \quad @_2(x, y) \leq \overline{@_2}(x, y).$$

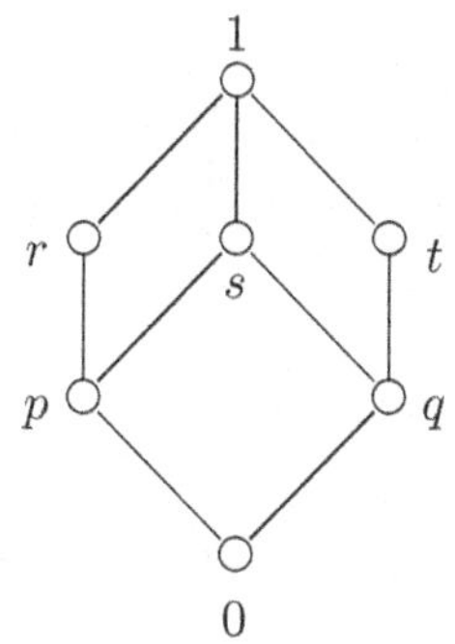

Fig. 1 Bounded lattice $(L, \preceq)$

Table 1 Agg. operator $@_2(x, y)$.

$@_2$	0	p	q	r	s	t	1
0	0	s	q	1	s	t	1
p	p	s	1	1	s	1	1
q	q	s	q	1	s	t	1
r	r	1	1	1	1	1	1
s	s	s	s	1	s	1	1
t	t	1	t	1	1	t	1
1	1	1	1	1	1	1	1

The sufficient conditions for the family $@_i$, such that the n-ary operator $A_@$ obtained by Proposition 2 produces (based on Definition 2) a family of binary operators identical to the ones from the starting family.

Proposition 4 *Let* $\{@_i : L^2 \rightarrow L \mid i \in \{1, \ldots, n-1\}\}$ *be a family of monotone mappings such that* $@_i(x, 0) = 0$, *and* $x \vee y \leq @_i(x, y)$, *for all* $i \in \{1, \ldots, n-1\}$, *and* $x, y \in L$, *and the next chain of inequalities* $@_1(x, y) \leq @_2(x, y) \leq \cdots \leq @_{n-1}(x, y)$ *holds. Let* $A_@ : L^n \longrightarrow L$ *be a function defined as*

$$A_@(x_1, x_2, \ldots, x_n) = @_{n-1}(@_{n-2}(\ldots @_2(@_1(x_1, x_2), x_3), \ldots, x_{n-1}), x_n)$$

and $\{\overline{@_i} : L^2 \rightarrow L \mid i \in \{1, \ldots, n-1\}\}$ *be the family generated by* $A_@$ *as in Definition 2. Then,* $@_k(x, y) = \overline{@_k}(x, y)$, *for all* $k \in \{1, \ldots, n-1\}$, *iff* $@_k(x, y) = @_k(@_{k-1}(x, y), y))$, *for all* $k \in \{2, \ldots, n-1\}$.

The following example of a finite bounded lattice and the related aggregation operators nicely illustrates the previous results.

Example 1 Let $L = \{0, p, q, r, s, t, 1\}$ be an underlying set of lattice (Fig. 1).

Given $@_1(x, y) = x \vee y$, and $@_2(x, y)$, $@_3(x, y)$ defined in the following Tables 1 and 2.

Table 2 Agg. operator $@_3(x, y)$.

$@_3$	0	p	q	r	s	t	1
0	0	s	s	1	1	1	1
p	p	s	1	1	1	1	1
q	q	s	s	1	1	1	1
r	r	1	1	1	1	1	1
s	s	s	s	1	1	1	1
t	t	1	1	1	1	1	1
1	1	1	1	1	1	1	1

It can be easily checked that the operators are monotone in all of the arguments, and that $x \vee y \preceq @_i(x, y)$ for all $i \in \{1, 2, 3\}$. Therefore, an n-ary operator $A_@$ can be constructed as

$$A_@(x, y, z, t) = @_3(@_2(@_1(x, y), z), t).$$

That is, if we need the joint value for e.g. 0, p, q, and r, and in that particular order, the calculation is as follows

$$\begin{aligned} A_@(0, p, q, r) &= @_3(@_2(@_1(0, p), q), r) \\ &= @_3(@_2(0 \vee p, q), r) \\ &= @_3(@_2(p, q), r) \\ &= @_3(1, r) \\ &= 1. \end{aligned}$$

The following example, based on the method proposed in [5], shows that the lattice approach can have two possible applications. Lattices can be used not only for calculating the interaction measure of segments of the investigated data, but in the previous step as well. Values assigned to membership functions that later on will form simple functions for all investigated data can be based on a lattice.

Example 2 Table 3 shows a very small subset of the Bilbao crime dataset concerning the physical and/or psychical gender violence in the city of Bilbao (Spain). More information is given in [5].

The membership functions μ_{DP}, μ_{DW}, μ_{DN}, and μ_{DR} are evaluated on the lattice $(L, \preceq)$, depicted in the previous example, in which, incomparable elements are required. For example, for expressing the level of danger of some of the neighborhoods, when the expert considers that the levels of danger of *teacher center* and *train station* are different although (s)he cannot say which one is the most dangerous. Moreover, (s)he cannot order the danger of some neighborhoods.

Also, as already mentioned, a lattice based aggregation can be used for calculating the interaction measure. When observing the Bilbao crime dataset from Table 3., measures of importance for each segment are $\nu(P)$, $\nu(D)$, $\nu(N)$, and $\nu(R)$, and are

Table 3 Bilbao crime dataset related to physical and/or psychical gender violence

	Place (P)	Day (W)	Neighborh. (N)	Rent (R)	μ_{DP}	μ_{DW}	μ_{DN}	μ_{DR}
Crime1	Flat	Working day	MATIKO	30952	1	p	p	p
Crime2	Flat	Weekend	OTXARKOAGA	20714	1	1	r	1
Crime3	Flat	Weekend	ERREKAL.	28638	1	1	t	p
Crime11	Teach. center	Working day	ATXURI	27881	q	p	s	p
Crime12	Train St.	Weekend	BASURTU	38947	p	1	q	q
Crime13	Urban P. Road	Weekend	ABANDO	57382	s	1	1	r

provided by an expert, while measures of interaction can be obtained by $A_@$, i.e.,

$$\nu(P, D) = A_@(\nu(P), \nu(D)) = @_1(\nu(P), \nu(D)),$$

$$\nu(P, D, N) = A_@(\nu(P), \nu(D), \nu(N)) = @_2(@_1(\nu(P), \nu(D)), \nu(N)),$$

$$\nu(P, D, N, R) = A_@(\nu(P), \nu(D), \nu(N), \nu(R))$$

$$= @_3(@_2(@_1(\nu(P), \nu(D)), \nu(N)), \nu(R)).$$

Hence, in this case, $A_@$ can be considered for obtaining a degree of relevance associated with each crime through both measure of interaction, and level of danger, and so give a final prioritization to be studied by the expert. This approach is of a special interest for the further development since it allows us to consider the incomparable elements, i.e., to model real life situations when a decision maker cannot express his or her preferences clearly.

4 Conclusions and Future Work

This paper complements the results introduced in [5] including the possibility of considering bounded lattices, which offers an extra level of flexibility in the applications. For example, it is possible to use intervals of values instead of the unit interval, or other general lattices as Example 1 takes into account. Further research would be an investigation of Choquet integral based on lattice valued monotone set functions as a tool for ranking forensics datasets. This new lattice based model will extend the applicability of the original model to missing date problems, e.g., working with metadata with some unavailable segments of data due to corruption of files.

References

1. Benvenuti, P., Mesiar, R., Vivina, D.: Monotone set functions-based integrals. In: Pap, E. (ed.) Handbook of Measure Theory, pp. 1329–1379. Elsevier, Amsterdam (2002)
2. Beliakov, G., Pradera, A., Calvo, T.: Aggregation Functions: A Guide for Practitioners. Springer, Berlin (2007)
3. Bustince, H., Fernandez, J., Mesiar, R., Montero, J., Orduna, R.: Nonlinear analysis: theory, overlap functions. Methods Appl. **72**(3–4), 1488–1499 (2010)
4. Choquet, G.: Theory of capacities. Annales de l'Institut Fourier **5**, 131–295 (1953)
5. Cornejo, M.E., Medina, J., Štajner-Papuga, I., Tepavčević, A.: On Choquet integral in ranking crimes. In: Cornejo, M.E., Harmati, I.A., Koczy, L.T., Medina-Moreno, J. (eds.) Computational Intelligence and Mathematics for Tackling Complex Problems 4. Studies in Computational Intelligence, vol. 1040, pp. 181–187. Springer, Cham (2023)
6. Cornejo, M.E., Medina, J., Štajner-Papuga, I., Tepavčević, A.: Choquet Integral in Ranking Crimes—On the Construction of a Measure of Interaction. submitted
7. Grabisch, M., Marichal, J., Mesiar, R., Pap, E.: Aggregations Functions. Cambridge University Press (2009)
8. Grabisch, M., Nguyen, H.T., Walker, E.A.: Fundamentals of Uncertainty Calculi with Applications to Fuzzy Inference. Kluwer Academics Publishers, Dordrecht (1995)
9. Karaçal, F., Mesiar, R.: Aggregation functions on bounded lattices. Int. J. General Syst. **46**:1, 37-51 (2017)
10. Koczy, L.T., Cornejo, M.E., Medina, J.: Algebraic structure of fuzzy signatures. Fuzzy Sets Syst. **418**, 25–50 (2021)
11. Lin, S.: Rank aggregation methods. WIREs Comp. Stat. **2**, 555–570 (2010). https://doi.org/10.1002/wics.111
12. Santos, H., Dimuro, G.P., Asmus, T.C., et al.: General grouping functions. Inf. Process. Manag. Uncertain. Knowl.-Based Syst. **1238**, 481–495 (2020)
13. Sugeno, M.: Theory of fuzzy integrals and its application. Doctoral thesis. Tokyo Institute of Technology (1974)
14. Qiao, J.: Overlap and grouping functions on complete lattices. Inf. Sci. **542**, 406–424 (2021)
15. Takači, A., Štajner-Papuga, I., Drakulić, D., Marić, M.: An extension of maximal covering location problem based on the choquet integral. Acta Polytechnica Hungarica **13**(4), 205–220 (2016)
16. Wang, Z., Klir, G.J.: Generalized Measure Theory. Springer, Berlin (2000)
17. Yu, P.L.H., Gu, J., Xu, H.: Analysis of ranking data. WIREs Comput. Stat. **11**, e1483 (2019). https://doi.org/10.1002/wics.1483
18. Zadeh, L.A.: Fuzzy sets. Inf. Control **8**, 338–353 (1965)

Power Consumption-Based Identification of Used Encryption Solution in IoT Environments

Aleksandar Jevremovic and Zona Kostic

1 Introduction

Computationally breaking the state-of-the-art encryption algorithms computationally is (still) virtually impossible. However, there are other types of attacks on data and systems that are using these algorithms. For example, Charles V. Wright et al. showed [7] that it is possible to extract words or phrases from encrypted VoIP calls. Michael et al. [2] proposed an approach that proved to be able to correctly identify an unknown speaker in 48% of cases in encrypted voice streams. In 2014, Jiangmin and Chan-Tin presented a method to identify Web browsers in encrypted communications [8]. Furat Al-Obaidy et al. used classification-based ML for identifying different social media applications in encrypted traffic [1].

Power consumption analysis is another method that could be used to extract some information from encrypted communications. For example, Amir Moradi et al. used a power analysis attack to extract keys used by the DES module on Xilinx Virtex-II FPGAs [5]. A similar attack, on the AES algorithm and ATMega328 Microcontroller [3] was presented in 2022 by Utsav Banerjee et al. and in 2015 on GPU-based AES [4].

In this paper, we present an approach to using power consumption (i.e. energy efficiency) to identify used protocol and encryption solutions in network communications. The method is oriented toward EdgeAI/IoT devices—Nvidia Jetson Nano, TX2, and AGX Xavier. Two cases are analyzed, when the device model is not known to the attacker, and when it is. We analyzed plain FTP transfer, OpenSSL, WireGuard, and OpenSSL and WireGuard combined. The results show that the proposed method

A. Jevremovic (✉)
Faculty of Informatics and Computing, Belgrade, Serbia
e-mail: ajevremovic@ieee.org

Z. Kostic
Harvard University, Cambridge, MA, USA
e-mail: zonakostic@seas.harvard.edu

M. E. Cornejo et al. (eds.), *Computational Intelligence and Mathematics for Tackling Complex Problems 5*, Studies in Computational Intelligence 1127,
https://doi.org/10.1007/978-3-031-46979-4_15

Table 1 Devices' hardware characteristics and configurations

Device	CPU	BogoMIPS	Memory
Jetson nano	Quad-core ARM cortex-A57	38.40	4 GB
Jetson TX2	Dual-core NVIDIA Denver2 + Quad-core ARM cortex-A57	62.50	8 GB
Jetson AGX Xavier	Octal-core NVIDIA carmel ARMv8.2 @ 2.26 GHz	62.50	32 GB

is reliable in determining if the encryption was used at all and that there is relatively high reliability in identifying the used encryption solution.

2 Methodology

In this research, we used three IoT devices—Jetson devices produced by Nvidia: Nano, TX2, and AGX Xavier. Additionally, Nano and AGX Xavier devices were tested in two different power modes—one that provides the highest performance, and one that provides the best energy efficiency. The reason we used different devices is that we wanted also to see if the results are transferable to different platforms or not. The hardware configuration of these devices is given in Table 1.

We assume that the content of the data is not known to the attacker, but the size of the data (10 GB) is. This assumption is not unrealistic in the way that many IoT installations share the same software applications that exchange fixed-size messages or messages whose size is not significantly variable. On the other hand, 10-gigabyte large streams are highly untypical for IoT applications, although in some cases there are continuous streams that consume full channel bandwidth. The reason why we used transfers like this in the experiment is that only during the longer periods of intensive CPU usage (e.g. for data encryption purposes) do devices switch to higher power modes where the power consumption is increased.

Another set up in the experiment is the use of 1 Gb/s wired communication channels. This is also untypical for IoT environments, where lower-speed wireless communications technologies are commonly used. The reason we used this setup is to eliminate the communication channel as a bottleneck for reaching maximum performance and power consumption. Also, this configuration is possible in setups where an IoT device plays a VPN (Virtual Private Network) gateway, positioned between the IoT environment and remote location.

We measured power consumption used by the transfer protocol/encryption solution by the difference between power consumption in the idle state and the power consumption during the transfer. The power consumption is expressed as energy

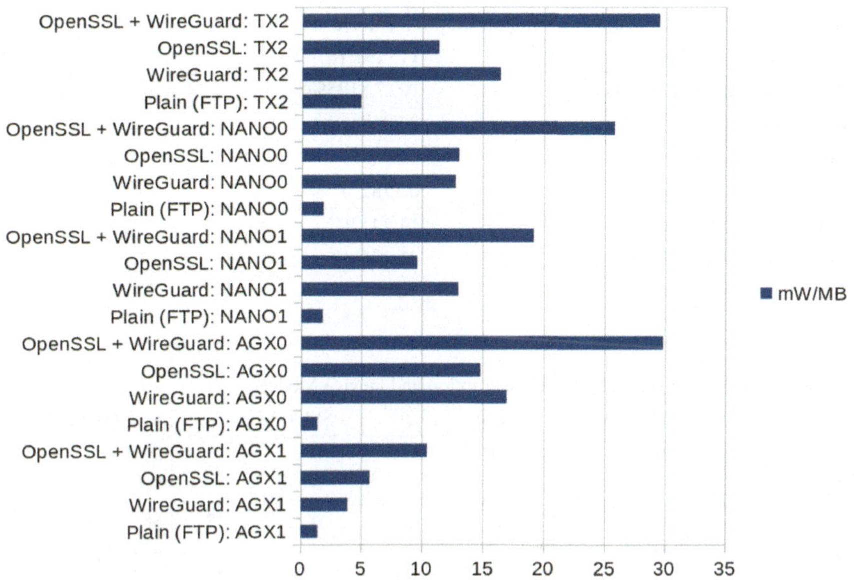

Fig. 1 Average power consumption when sending data, per device/protocol

efficiency, or milliwatts consumed per megabyte of data transferred (mW/MB). This efficiency is shown on the chart in Fig. 1.

For the protocol identification, we used the Random Forest algorithm, with a 10-fold cross-validation test mode. The data was collected within 128 sessions, i.e. four repeated sessions per device/mode/protocol.

3 Results

The main results are presented in Tables 2 and 4. The data in Table 2 shows that, when both power consumption and device model are used, the classification of the used transfer protocol was correct for 118 sessions (92%).

The confusion matrix for this classification (Table 3) shows that there were no errors in classification when plain FTP was used, or when both SSL and WireGuard were used. This is rather expected since these two cases are extremes in power consumption. However, the classification algorithm confused WireGuard and SSL ten times in 64 sessions. This is expected since these two protocols have similar power consumption (e.g. almost identical on the Nano device with power mode 0).

The results when only power consumption information was used (i.e. the device model is not known) are shown in Tables 4 and 5. The classification was expectedly less correct (82%, compared to 92% when the device model was known).

Table 2 Classification results (Power Consumption and Device Model)

Correctly classified instances	118 (92.1875%)
Incorrectly classified instances	10 (7.8125%)
Kappa statistic	0.8958
Mean absolute error	0.0542
Root mean squared error	0.1782
Relative absolute error	14.4453%
Root relative squared error	41.1356%
Total number of instances	128

Table 3 Confusion matrix (Power Consumption and Device Model)

FTP	WG	SSL	SSL+WG	← Classified as
32	0	0	0	FTP
0	27	5	0	WG
0	5	27	0	SSL
0	0	0	32	SSL+WG

Table 4 Classification results (Power Consumption Only)

Correctly classified instances	105 (82.0313%)
Incorrectly classified instances	10 (17.9688%)
Kappa statistic	0.7604
Mean absolute error	0.1
Root mean squared error	0.2561
Relative absolute error	26.6527%
Root relative squared error	59.1154%
Total number of instances	128

The confusion matrix for this classification shows that, also, there was no confusion between plain FTP and encrypted transfers. However, in this case, the SSL+WireGuard combination was classified as SSL or WireGuard only encryption in five cases. This can be explained by the different energy efficiency of different devices.

In both cases, there was no confusion between plain and encrypted data transfer. This suggests that the proposed method can be used for reliable distinction (at least, within the constraints we used in the experiment) if some encryption solution was used, or the transfer was done by using the plain FTP protocol.

Table 5 Confusion matrix (Power Consumption Only)

FTP	WG	SSL	SSL+WG	← Classified as
32	0	0	0	FTP
0	21	8	3	WG
0	4	25	3	SSL
0	3	2	27	SSL+WG

4 Conclusion

In this paper, we presented an approach to identifying used protocol and encryption solutions in network communications, based on the power consumption or energy efficiency. We targeted EdgeAI/IoT environment and analyzed different Nvidia Jetson devices, in different power modes, and different protocols: plain FTP, OpenSSL, WireGuard and OpenSSL+WireGuard combined.

The results show that the proposed approach is reliable in detecting if encryption is used at all, and somewhat reliable in identifying used encryption solutions. Some assumptions in the experiment are not particularly common in IoT environments, like using wired 1 Gb/s links and transfers, or using the device's CPU for encryption purposes only. However, the results show the potential of the presented approach, and future work includes detailed profiling and training with more common scenarios and data, as well as using more advanced classification algorithms.

References

1. Al-Obaidy, F., Momtahen, S., Hossain, M.F., Mohammadi, F.: Encrypted traffic classification based ml for identifying different social media applications. In: 2019 IEEE Canadian Conference of Electrical and Computer Engineering (CCECE), pp. 1–5 (2019)
2. Backes, M., Doychev, G., Dürmuth, M., Köpf, B.: Speaker recognition in encrypted voice streams. In: Gritzalis, D., Preneel, B., Theoharidou, M. (eds.) Computer Security—ESORICS 2010, pp. 508–523. Springer, Berlin (2010)
3. Banerjee, U., Ho, L., Koppula, S.: Power-based side-channel attack for aes key extraction on the atmega328 microcontroller (2022)
4. Luo, C., Fei, Y., Luo, P., Mukherjee, S., Kaeli, D.: Side-channel power analysis of a gpu aes implementation. In: 2015 33rd IEEE International Conference on Computer Design (ICCD), pp. 281–288 (2015)
5. Moradi, A., Barenghi, A., Kasper, T., Paar, C.: On the vulnerability of fpga bitstream encryption against power analysis attacks: Extracting keys from xilinx virtex-ii fpgas. In: Proceedings of the 18th ACM Conference on Computer and Communications Security, CCS '11, pp. 111–124. Association for Computing Machinery, New York (2011)

6. Saputra, H., Vijaykrishnan, N., Kandemir, M., Irwin, M.J., Brooks, R., Kim, S., Zhang, W.: Masking the energy behavior of des encryption [smart cards]. In: 2003 Design, Automation and Test in Europe Conference and Exhibition, pp. 84–89. IEEE (2003)
7. Wright, C.V., Ballard, L., Coull, S.E., Monrose, F., Masson, G.M.: Uncovering spoken phrases in encrypted voice over ip conversations. **13**(4) (2010)
8. yu, J., Chan-Tin, E.: Identifying webbrowsers in encrypted communications. In: Proceedings of the 13th Workshop on Privacy in the Electronic Society, WPES '14, pp. 135–138. Association for Computing Machinery, New York

Maritime Cyber-Attacks Detection Based on a Convolutional Neural Network

Jamal Raiyn

1 Introduction

Over the last few years, various geo-positioning technologies have been used in maritime transportation systems to estimate the location of vessels, such as satellite-positioning technologies (i.e. GNSS and GPS), wi-fi positioning systems, and cellular positioning systems. To improve the estimation of maritime locations, a GNSS-based positioning system for maritime use was considered, which has the advantages of accuracy and low processing time. The basic operating principle of satellite navigation systems is to calculate a user's position from a GNSS signal. However, the quality of raw GNSS measurements (also called observables) is affected by several factors, associated with satellites, signal propagation, and receivers. The signal transmitted by a satellite propagates through the atmosphere, where it is subject to delays caused by ionospheric and tropospheric media. At ground level, a multipath effect can occur (namely, the reception of signals reflected from obstacles such as buildings surrounding the receiver), causing one of the largest types of errors, a type that is difficult to model because the errors depend heavily on the receiver environment. The influence of obstacles is most significant when a GNSS signal from a satellite arrives at the receiver only indirectly, since it is much more difficult to discard the faulty measurement. This type of phenomenon is described *non-line-of-sight*" (NLOS).

Maritime vessels use various technologies to establish communication with ports and other vessels [1], However, they are susceptible to assaults by cyber-attackers, who take advantage of a situation when systems are easily accessible. If a cyber-attacker discovers a weakness in a certain type of vessel or in a company's electronic system, a lack of information security can lead to criminal and terrorist acts, which could eventually cost lives. The most common security schemes against such attacks

J. Raiyn (✉)
Computer Science Department, Al Qasemi Academic College, Baqa Al Gharbiah, Israel
e-mail: raiyn@qsm.ac.il

M. E. Cornejo et al. (eds.), *Computational Intelligence and Mathematics for Tackling Complex Problems 5*, Studies in Computational Intelligence 1127,
https://doi.org/10.1007/978-3-031-46979-4_16

are based on cryptographic approaches, which are the most often used approaches in communication technologies [6]. The main goal of an encryption system is to disguise classified messages and render them unreadable to all unauthorized persons. Encryption cannot prevent sent files from getting to unauthorized persons, but it can prevent recipients from understanding their contents. The goal of this research project is to protect data from cyber attackers who aim to alter the course of maritime vessels and increase port and traffic congestion. To accomplish this, a convolutional neural network (CNN) is proposed [11].

The aims of the system are, first, to detect anomalies in the data received by ships, and second, to extract the features of GNSS data that make it possible to identify sources of anomalous data. By applying CNNs, we can recognize types of ships and their associated types of cyber-attacks, and using accurate measurement methods, such as root-mean square error (RMSE) and mean absolute error (MAE), we can make comparisons that reveal anomalies. In this paper, we analyze the results and identify the best methods for prediction.

2 Cybersecurity in Maritime Transportation Systems

Cyber-crime and cyber-terrorism incidents have been increasing exponentially. The main task of maritime cybersecurity is to make maritime traffic secure from threats [2]. This involves port security, maritime cybersecurity and cyber-attack detection. This paper focuses on maritime cyber-attack detection [2, 3]. Cyber-attacks are hostile actions that attempt to bypass the security mechanisms of computer systems. Cyber-attack detection has been defined as "the problem of identifying individuals who are using a computer system without authorization and those who have legitimate access to the system but are abusing their privileges. We add to this definition the identification of attempts to use a computer system without authorization or to abuse existing privileges. Identifying abnormal behavior on a host or network is done with anomaly detectors. They function on the assumption that attacks leave signs that differ from those of legitimate activity and can therefore be detected by systems that can identify these differences. Using statistical methods to detect anomalies is one of the oldest techniques applied in IDS research. In this approach, normal user behavior is first defined based on what is acceptable within usage policies of a system.

3 System Model

This section gives an overview of scenarios of cyber-attacks on maritime transportation systems [4, 5] and focuses on cyber-attack detection in maritime systems, using augmented GNSS. Augmented GNSS is a cyber-immune system for analyzing and detecting anomalous and possibly malicious activity. The design concept is to create a self-immunizing system through anomaly-based statistical measurements. Cyber security in the augmented GNSS is based on comparing stored and newly collected data. When a vessel receives the wrong data, the control unit takes immediate action, such as sending an alarm. The novelty of the proposed approach lies in building in a self-immunization system and adding new components to detect anomalous data [6]. It provides enhanced security based on a CNN.

3.1 Convolutional Neural Network 2D

The proposed system for maritime transportation uses a convolutional neural network (CNN) to detect anomalous data [10]. The data input is powered by online map services that include information on various types of ships, such as cargo ships, tankers, passenger ships, fishing vessels, high speed craft, tugs and yachts as illustrated in Fig. 1. The figure shows an example of a raw snapshot of a maritime transportation system, the Suez Canal, whose configuration is narrow and crooked. The snapshot was acquired through a free API an online map provider; it shows a regular day of canal operations, June 2, 2022. Snapshots like this were the initial source of marine traffic congestion data for the projects.

Each raw snapshot was 400 pixels wide and 400 pixels high. The procedure consisted of cleaning the data, extracting the attributes, and structuring the data as illustrated in Fig. 2.

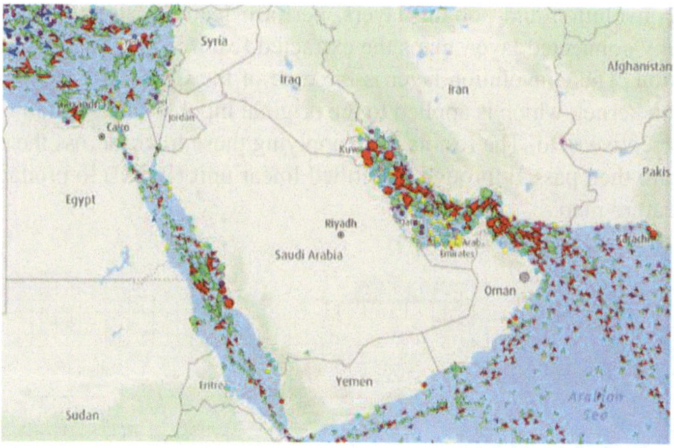

Fig. 1 Vessel classes

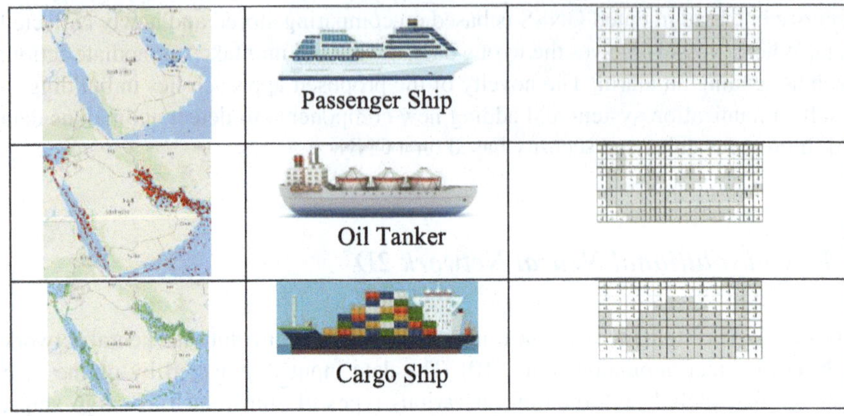

Fig. 2 Feature extraction

3.2 Oil Tanker Recognition

Ships that transport energy resources are a major target of cyber-attackers. In this study, a CNN was applied to identify vessel types and their associated types of cyber-attack. A CNN is designed to automatically and adaptively learn the spatial hierarchy of features, from low-to high-level patterns. It is typically composed of three layers: convolution, pooling, and fully connected layers as illustrated in Fig. 3. The first two, the convolution and pooling layers, perform feature extraction, whereas the third, a fully connected layer, maps the extracted features into final output, such as classification. The convolution layer is the core of the CNN. The weights define a convolution kernel, which is applied to the original input, a small window at a time, called a receptive field. The results from applying these filters across the entirety of the input are then passed through a rectified linear unit (ReLU) to produce what is called a feature map.

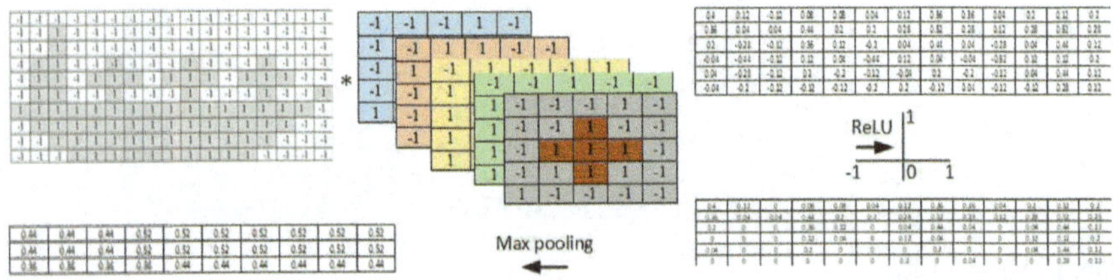

Fig. 3 CNN architecture

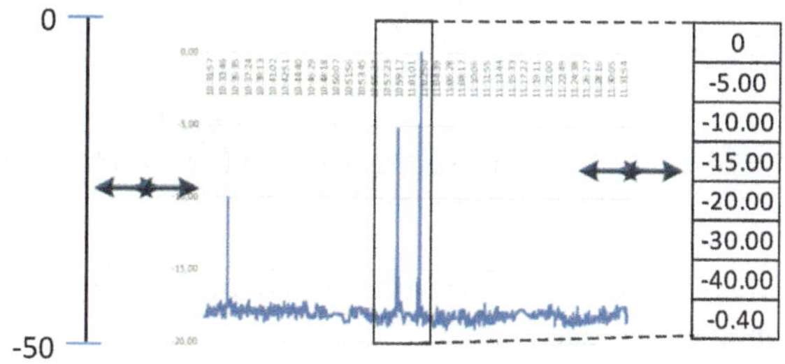

Fig. 4 Anomalous data detection

3.3 *Anomalous Data Detection*

A CNN is a type of deep learning model for processing data, one dimensional, two dimensional and three dimensional. The first dimension of the CNN is used to detect anomalous data received by ships, as illustrated in Fig. 4.

- Input layer

The input of the model consists of GNSS data received by ships. The lengths of the input and output time intervals can be expressed as F and P, respectively. The model input can be written as:

$$x^i = [m_i, m_{i+1}, \ldots, m_{i+P-1}], i \in [1, N - P - F + 1]$$

where I is the sample index, N is the length of a time interval, and m_i is a column vector representing raw GNSS data.

The interference power range for the "sensitive method" is defined as (dBm) [0: – 50]. The CNN is applied to detect anomalous GNSS data in ships. The extraction of GNSS data features is achieved through the combined operations of the convolution and pooling layers. The output from the first convolution and pooling layers can be written as:

$$o_1^j = pool(\sigma(W_1^j x_1^j + b_1^j)), j \in [1, c_1] \tag{1}$$

where σ is the activation function.

In the prediction, the features learned and output by traffic feature extraction are concatenated into a dense vector that contains the final and highest-level features of the transportation network input. This dense vector can be written as:

$$o_L^{fallten} = flatten([o_L^1, o_L^2, \ldots, o_L^j]), j = c_L \tag{2}$$

where L is the depth of CNN.

Finally, the vector is transformed into output through the fully connected layer. The output can be written as

$$\tilde{y} = W_f o_L^{flatten} + b \tag{3}$$

$$= W_f(fallten(pool(\sigma\left(\sum_{k=1}^{c_L-1}(W_L^j x_L^k + b_L^j))))\right) + b_f \tag{4}$$

where W_f and b_f are the parameters of the fully connected layer, and $\tilde{y}$ is the predicted anomalous data.

The CNN applies convolutional filters to its input layer and obtains local connections only where local input neurons are connected to an output neuron (in the convolution layer). Hundreds of filters are sometimes applied to the input and the results are merged in each layer. Since one filter can extract one GNSS data feature from the input layer, hundreds of filters can extract hundreds of GNSS data features.

To secure the maritime communication is used an encryption approach. The main goal of any encryption system is to disguise the classified message and render it unreadable to all unauthorized persons. Cryptographic systems are most often used in communication technologies. Encryption cannot prevent sent files from getting to the unauthorized person, but they can prevent the recipient from understanding their contents. To secure the received data by ship-to-satellite communication, two phases are performed. The first phase, it is considered GNSS raw data. The CNN detects anomalies data based on augmented GNSS. The augmented GNSS has been designed as a cyber -immune system for analyzing and detecting anomalous activities and their relation to malicious activities. The cyber security in the augmented GNSS aims to secure stored and newly collected data. The second phased used a blockchain strategy to secure the data communication between maritime device Internet of Things devices [9]. Maritime Blockchain approach [7, 8] stored the data in a way that makes it difficult for unauthorized user to change, manipulate, or delete data. Maritime Blochchain technology is advanced to include cyber agent that monitor the data communication and the source of the data as illustrated in Fig. 5.

To keep data integrity, reliability, and accuracy, a CNN is used to detect anomalies data. Anomaly detectors identify abnormal unusual behavior on a host or network. They function on the assumption that attacks are different from legitimate activity and can therefore be detected by systems that identify these differences. A statistical analysis is the most used to detect anomalies data. The loss function was a customized mean squared error (MSE) defined in Eq. (5).

$$MSE = \frac{1}{n}\sum_{i=1}^{n}(y_i - \tilde{y}_i)^2 \tag{5}$$

The cost function is a special type of functions, which help us minimize the error, and reach as close as possible to the expected output. Cost function uses two parameters to calculate the error, the first parameter is the estimate output of the CNN model (also called the prediction) and the second one is the actual output. The mean squared error is considered one of the most familiar loss functions as it is really

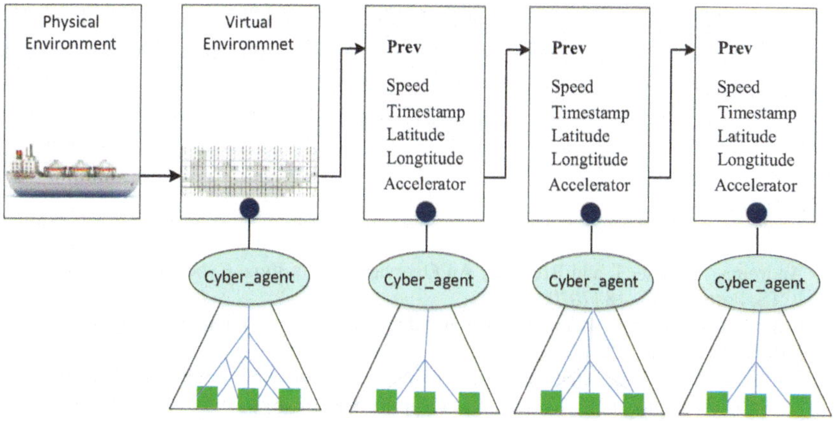

Fig. 5 Blockchain strategy

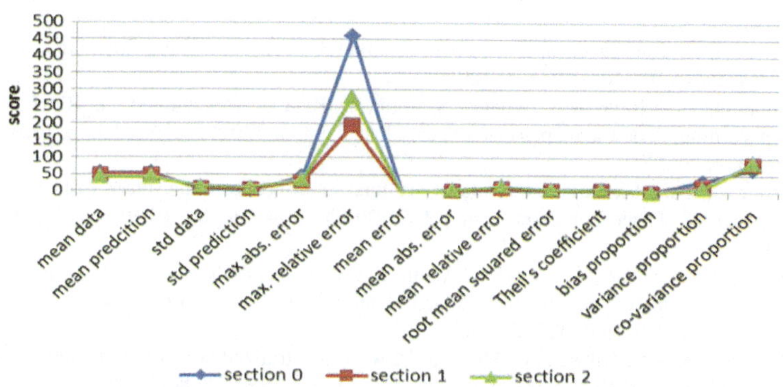

Fig. 6 Statistical measurment error

like the least square loss function. It directly calculates the difference between the predicted result and the true label. To achieve a good performance of the CNN, the training dataset should be free from anomalies. An exponential smoothing scheme used statistical functions to detect anomalies data. The simple statistical functions detect the univariate anomalies feature values in data sets as illustrated in Fig. 6.

4 Conclusion

The increasing integration of maritime IT systems with global communications systems means that the maritime domain is now part of cyberspace. This paper addresses maritime cybersecurity. A CNN is proposed to identify ship types and to detect anomalous GNSS data received by ships. There are two types of anomalous data. One type is caused by geographical factors, the other by cyber attackers. The CNN is designed to detect anomalous data and identify ships that are being targeted by cyber-attackers.

References

1. Androjna, A., Srse, J. and Brcjo, T.: An Overview of Maritime Cyber Security Challenges, ICTS, Portoroz, 17–18 (2020)
2. IMO: Maritime cyber risk management in safety management system. Resolut. MSC **428**(98) (2017). Accessed 16 June
3. Lo, C.: GPS spoofing: what's the risk for ship navigation? Ship Technology Global (2019). https://www.shiptechnology.com/features/ship-navigation-risks/. Accessed 15 July
4. Boyes, H.: Maritime cyber security—securing the digital seaways. Eng. Technol. Ref. (2014)
5. Zalewski, P.: Real-time GNSS spoofing detection in maritime code receivers. Sci. J. Maritime Univ. Szczecin **38**(110), 118–124 (2014)
6. Raiyn, J.: A survey of cyber attack detection strategies. Int. J. Secur. Appl. **8**(1), 247–256 (2014)
7. Berman, D.S., Buczak, A.L., Chavis, J.S., Cherita L.C.: A survey of deep learning methods for cyber security. Inf. **10**(122), 1–35 (2019)
8. Muhammad, S., Raja, W.A., Khaled, S., Ibrar, Y., Raja, J., Mohammed, O.: Blockchain for deep learning: review and open challenges. Clust. Comput. (2021). https://doi.org/10.1007/s10586-022-03582-7
9. Kuzlu, M., Fair, C., Guler, O.: Role of artificial intelligence in the internet of things (IoT) cybersecurity. Discov. Internet Things **1**, 7 (2021). https://doi.org/10.1007/s43926-020-00001-4
10. Saleem, M.A., Senan, N., Wahid, F., Aamir, M., Samad, A., Khan, M.: Comparative analysis of recent architecture of convolutional neural network. Hindawi Math. Prob. Eng. 1–9 (2022). https://doi.org/10.1155/2022/7313612
11. Abdulrazzaq, S.T., Omar, F.S., Mustafa, M.A.: Decentralized security and data integrity of blockchain using deep learning techniques. Period. Eng. Nat. Sci. **8**(3), 1911–1923 (2020)

Epistemic Logic and Theory of Mind for Modelling Group Dynamics of Criminal Organizations

Stefania Costantini, Andrea Formisano, and Valentina Pitoni

1 Introduction

Agents and Multi-Agent Systems (MAS) are a suitable and widely used tool in Artificial Intelligence (AI), to model societies whose members are to some extent cooperative towards each other. Notice that, a criminal organization can in fact be seen as a group of intelligent agents aiming to cooperate in order to reach common objectives.

The Logic of "Inferable" *L-DINF* has been especially devised to model the dynamics of groups of cooperative agents. In this paper, we show how this agent-oriented logical framework [1, 2] allows a designer to formalize the capability to construct and execute joint plans within a group of agents, that is which happens in criminal organizations. Proofs in our logical system may allow investigators to understand if and how such a group might reach its objectives. Explainability is a crucial factor, as it may allow investigators to understand the system's conclusions; the syntax of our logic has been deliberately developed in view of generating human-level explanations.

The *L-DINF* logic allows one to model the fact that agents belong to groups. It also supports reasoning on actions' *costs* and agents' *budgets*, where an agent able to perform an action but not owning the needed budget can be supported by its group to cover the cost. Agents groups take into consideration agents' preferences on actions'

S. Costantini · V. Pitoni
DISIM, Università di L'Aquila, L'Aquila, Italy
e-mail: stefania.costantini@univaq.it

V. Pitoni
e-mail: valentina.pitoni@univaq.it

A. Formisano (✉)
DMIF, Università di Udine, Udine, Italy
e-mail: andrea.formisano@uniud.it

M. E. Cornejo et al. (eds.), *Computational Intelligence and Mathematics for Tackling Complex Problems 5*, Studies in Computational Intelligence 1127,
https://doi.org/10.1007/978-3-031-46979-4_17

execution, and agents' *roles* (in terms of the actions that each agent is allowed to do in the context of its group). In addition, normally an action can be performed only by an agent which is able and allowed to perform it; but, in exceptional circumstances, for lack of alternatives, also an agent that is able, although not allowed, can perform it.

There are classes of applications where it can be useful to represent group dynamics, and to understand the behaviour of others; i.e., agents should be able to assess or hypothesize what other agents believe and intend to do. This accounts to be able to represent aspects of "*Theory of Mind*", which is the set of social-cognitive skills involving: the ability to attribute mental states, including desires, beliefs, and knowledge, to oneself and to other agents; and, importantly, the ability to reason about the practical consequences of such mental states. Such ability is crucial for prediction and interpretation of other agents' behavioural responses [3]. This feature might be particularly important for investigators, in order, for instance, to understand the behaviour of a suspect. In ongoing work [4] we have devised an extension of *L-DINF* able to represent aspects of Theory of Mind.

All these features are defined in a modular way, so as to be composed to model different group behaviours. We introduced to this aim special functions in the definitions of canonical models (not shown here for lack of space, cf., e.g., [5], where we also discuss the relationship of logic *L-DINF* with related work). For the practical realization of such functions, we envisage separate modules from which agent's logical theory "inputs" the results. Such modules might be specified even in some other logic or also, pragmatically, via pieces of code.

The paper is organized as follows. In Sect. 2 we introduce, reporting material from previous papers, syntax and semantics of *L-DINF*. Section 3, we discuss a significant example of application of the new logic, concerning how to represent planning activities in criminal organizations. Finally, in Sect. 4 we conclude.

2 Logical Framework

L-DINF is a logic including a static component and a dynamic one. The former, called *L-INF*, is a logic of explicit beliefs and background knowledge. The latter defines dynamic operators to capture consequences of agents' inferential actions, on their beliefs. Formulas of *L-DINF* are built out of a set $Atm = \{p, q, \ldots\}$ of atomic propositions and a set Atm_A representing the physical actions that agents can perform. Let $Agt = \{1, 2, \ldots, n\}$ be a set of integer numbers identifying n agents. The language $\mathcal{L}_{L\text{-}DINF}$ of *L-DINF*, is defined by the following grammar [1, 5]:

$$
\begin{aligned}
\varphi, \psi &::= p \mid \neg\varphi \mid \varphi \wedge \psi \mid \mathbf{B}_i\, \varphi \mid \mathbf{K}_i\, \varphi \mid do_G(\phi_A) \mid do_G^P(\phi_A) \mid can\_do_G(\phi_A) \mid \\
&\qquad intend_G(\phi_A) \mid exec_G(\alpha) \mid [G{:}\alpha]\, \varphi \mid pref\_do_i(\phi_A, d) \mid pref\_do_G(i, \phi_A) \\
\alpha &::= \vdash(\varphi,\psi) \mid \cap(\varphi,\psi) \mid \downarrow(\varphi, \psi) \mid \dashv(\varphi, \psi)
\end{aligned}
$$

where $p \in Atm$, $i \in Agt$, $G \subseteq Agt$, $\phi_A \in Atm_A$, and $d \in \mathbb{N}$.[1] The language of *inferential actions* of type α is denoted by $\mathcal{L}_{\mathsf{ACT}}$. The static part *L-INF* of *L-DINF*, includes only those formulas not having sub-formulas of form $[G{:}\alpha]$. The modality $\mathbf{K}_i$ expresses background knowledge of agent i, while $\mathbf{B}_i$ is used to model the explicit beliefs j. We do not consider here formulas involving nesting of modalities. Hence, in formulas of the forms $\mathbf{B}_i\,\varphi$ and $\mathbf{K}_i\,\varphi$, the subformula φ does not involve any operator $\mathbf{B}_j$ and $\mathbf{K}_j$ (for any i, j). Unlike explicit beliefs, i.e., facts and rules acquired via perceptions during an agent's operation and kept in the *working memory*, an agent's background knowledge is assumed to satisfy *omniscience* principles, such as closure under conjunction and known implication, and closure under logical consequence, and introspection. Hence, $\mathbf{K}_i$ is actually the S5 modal operator often used to model/represent knowledge. The fact that background knowledge is closed under logical consequence is justified because we conceive it as a kind of stable reliable *knowledge base*, or *long-term memory*. We assume the background knowledge to include: facts (formulas) known by the agent from the beginning, and facts the agent has later decided to store in its long-term memory (by means of some decision mechanism not treated here) after having processed them in its working memory. We therefore assume background knowledge to be irrevocable, in the sense of being stable over time.

We consider the set of agents as partitioned in groups: each agent i always belongs to a single group $G \subseteq Agt$. Any agent i, at any time, can perform a (physical) action $join_A(i, j)$, to change its group and join j's one.

Let us briefly describe the intended informal meaning of basic formulas of *L-INF*. The expression $intend_G(\phi_A)$ indicates the intention of all agents in G to perform ϕ_A in the sense of the BDI agent model [6]. This intention can be part of agents' knowledge base from the beginning, or it can be derived later. The formula $do_G(\phi_A)$ indicates the *actual execution* of ϕ_A by an agent in G. Whenever an agent derives this formula the physical action ϕ_A is supposed to be performed via some kind of *semantic attachment* which links the agent to the external environment. Moreover, this event is automatically recorded by the new belief $do^P_G(\phi_A)$ (postfix "P" standing for "past" action). The do^P_G beliefs constitute a *history* of the agent's operation, so they might be useful for the agent to reason about its own past behaviour, and/or, importantly, they may be useful to provide *explanations* to human users.

The expression $can\_do_G(\phi_A)$ is to be seen as an enabling condition, indicating that at leas one agent $i \in G$ is enabled to execute action ϕ_A. The formula $pref\_do_i(\phi_A, d)$ indicates the level d of preference/willingness of agent i to perform ϕ_A, while the formula $pref\_do_G(i, \phi_A)$ indicates that agent i exhibits the *maximum level* of preference on performing ϕ_A among all members of its group G. If a group of agents intends to perform an action ϕ_A, this will entail that the entire group intends to do ϕ_A, that will be enabled to be actually executed only if at least one agent $i \in G$ can do it, i.e., it can derive $can\_do_i(\phi_A)$.

[1] Other Boolean operators are defined from $\neg$ and $\wedge$ in the standard manner. Moreover, whenever $G = \{i\}$ we will write i as subscript in place of $\{i\}$. So, for instance, we often write $do_i(i, \phi_A)$ instead of $do_{\{i\}}(i, \phi_A)$ and similarly for other constructs.

The formula $[G : \alpha]\,\varphi$, where α is an inferential action, states that "φ holds after α has been performed by one agent of G, and all agents in G have common knowledge about this fact". We distinguish four types of inferential actions α:

- $\downarrow(\varphi, \psi)$: it infers the belief ψ from the belief φ if, according to agent's background knowledge, ψ is a logical consequence of φ.
- $\cap(\varphi, \psi)$: it closes the explicit beliefs φ and ψ under conjunction.
- $\dashv(\varphi, \psi)$: it performs a simple form of "belief revision". It removes the belief ψ from the working memory if φ is believed and, according to agent's background knowledge, $\neg\psi$ is logical consequence of φ. Both ψ and φ are required to be atoms.
- $\vdash(\varphi, \psi)$: it adds ψ to the working memory in case φ is believed and, according to agent's working memory, ψ is logical consequence of φ.

The formula $exec_G(\alpha)$ expresses executability of inferential actions by a group G of agents. It has to be read as: "α is an inferential action that an agent in G can perform". To execute actions, agents may have to pay costs, so they must have a consistent budget available. Moreover, agents are entitled to perform only those physical actions that they conclude they can do. Agents belonging to a group are assumed to be cooperative. An action can be executed by a group if at least one agent in the group is able to execute it, and the group has the necessary budget available, sharing the cost according to some policy. The cooperative nature of the agents manifests itself also in selecting, among the agents that are able to do an action, the one(s) which best prefer to perform it.

Semantic of *L-INF* is specified in terms of *L-INF models*, as introduced by Definition 1 (see [1, 4] for a detailed description of semantics of the logic). Many relevant aspects of an agent's behaviour are specified in the definition of *L-INF model*, including which mental and physical actions an agent can perform, which is the cost of an action and which is the budget that the agent has available, which is the preference degree of the agent to perform each action. This choice has the advantage of keeping the complexity of the logic under control, and of making these aspects modularly modifiable.

Definition 1 A model is a tuple $M = (W, N, \mathcal{R}, E, B, C, A, H, P, V)$ where:

- W is a set of worlds (or situations);
- $\mathcal{R} = \{R_i\}_{i \in Agt}$ where for each i, $R_i \subseteq W \times W$ is an equivalence relations on W;
- $N : Agt \times W \longrightarrow 2^{2^W}$ is a neighborhood function such that, for each $i \in Agt$, each $w, v \in W$, and each $X \subseteq W$ these conditions hold:

  (C1) if $X \in N(i, w)$ then $X \subseteq \{v \in W \mid wR_iv\}$,
  (C2) if wR_iv then $N(i, w) = N(i, v)$;

- $E : Agt \times W \longrightarrow 2^{\mathcal{L}_{\mathsf{ACT}}}$ is an executability function of mental actions such that, for each $i \in Agt$ and $w, v \in W$, it holds that:

  (D1) if wR_iv then $E(i, w) = E(i, v)$;

- $B : Agt \times W \longrightarrow \mathbb{N}$ is a budget function such that, for each $i \in Agt$ and $w, v \in W$, the following holds

  (E1) if wR_iv then $B(i, w) = B(i, v)$;

- $C : Agt \times \mathcal{L}_{\mathsf{ACT}} \times W \longrightarrow \mathbb{N}$ is a cost function such that, for each $i \in Agt$, $\alpha \in \mathcal{L}_{\mathsf{ACT}}$, and $w, v \in W$, it holds that:

  (F1) if wR_iv then $C(i, \alpha, w) = C(i, \alpha, v)$;

- $A : Agt \times W \longrightarrow 2^{Atm_A}$ is an executability function for physical actions such that, for each $i \in Agt$ and $w, v \in W$, it holds that:

  (G1) if wR_iv then $A(i, w) = A(i, v)$;

- $H : Agt \times W \longrightarrow 2^{Atm_A}$ is an enabling function for physical actions such that, for each $i \in Agt$ and $w, v \in W$, it holds that:

  (G2) if wR_iv then $H(i, w) = H(i, v)$;

- $P : Agt \times W \times Atm_A \longrightarrow \mathbb{N}$: is a preference function for physical actions ϕ_A such that, for each $i \in Agt$ and $w, v \in W$, it holds that:

  (H1) if wR_iv then $P(i, w, \phi_A) = P(i, v, \phi_A)$;

- $V : W \longrightarrow 2^{Atm}$ is a valuation function.

Let us consider the set $R_i(w) = \{v \in W \mid wR_iv\}$, for $w \in W$. This set identifies the situations that agent i considers possible at world w. It is the *epistemic state* of agent i at w. In cognitive terms, $R_i(w)$ can be conceived as the set of all situations that agent i *can retrieve* from its long-term memory and reason about. While $R_i(w)$ concerns background knowledge, $N(i, w)$ is the set of all facts that agent i explicitly believes at world w, a fact being identified with a set of worlds. Hence, if $X \in N(i, w)$ then, the agent i has the fact X under the focus of its attention and believes it. We say that $N(i, w)$ is the explicit *belief set* of agent i at world w.

Executability of inferential actions is determined by the function E, i.e., $E(i, w)$ is the set of inferential actions that agent i can execute at world w. The value $B(i, w)$ is the budget i has available to perform inferential actions. $C(i, \alpha, w)$ is the cost to be paid by i to execute the inferential action α in w. $N(i, w)$ and $B(i, w)$ are updated whenever a mental action is performed (as described in [1]). The executability of physical actions is determined by tfunction A: $A(i, w)$ is the set of physical actions that i can execute at world w. $H(i, w)$ instead is the set of physical actions that agent i is enabled by its group to perform. Which means, H defines the *role* of an agent in its group, via the actions that it is allowed to execute. Agent's preference on executability of physical actions is determined by the function P: $P(i, w, \phi_A) \in \mathbb{N}$ indicates the degree of willingness of agent i to execute ϕ_A at world w. A notion of satisfaction for formulas of *L-INF*, w.r.t. a model M. Semantics of inferential actions is then defined in terms of changes the actions cause on the neighborhoods; namely, in terms of updates of the model, which correspond to updates in agents' working memory (cf. [1, 2]).

Modularity is supported in out approach, because functions A, P and H (i.e., executability of physical actions, preferences, and permissions) are not meant to be built-in. Rather, they can be defined via separate sub-theories, possibly defined using different logics, or, in a practical approach, even via pieces of code.

3 Example: Group Dynamics of Criminal Organizations

In this section, we propose a simple example illustrating how the logic framework outlined so far can be exploited to model, understand, and reason about dynamics of criminal organizations. Consider a group of three criminals/agents, that intend to cooperate to rob in a bank vault:

- agent 1 is able to dig a tunnel to the vault, but he needs a small excavator;
- agent 2 is able to open the security door, by means of suitable hardware tools;
- agent 3 will go to buy the needed tools, and will act as a guard outside the bank.

The group receives notification by some accomplice that is the right time to act (as, for instance, vigilance is reduced), so they decide to organize themselves for doing it. The group will reason, and devise the intention/goal $\mathbf{K}_i(intend_G(rob\_bank))$ (for $i = 1, 2, 3$). The group will now be required to perform a planning activity. Assume that, as a result of the planning phase, the knowledge base of each agent i contains the following rule, that specifies how to reach the intended goal in terms of actions to perform and sub-goals to achieve:

$$\begin{aligned}\mathbf{K}_i\big(intend_G(rob\_bank) \rightarrow\ & intend_G(buy\_material) \wedge \\ & intend_G(dig\_tunnel) \wedge \\ & intend_G(open\_security\_door)\big).\end{aligned}$$

For $i = 1, 2, 3$, the following consequences are entailed thanks to modal axioms of S5 logic, that, as mentioned, rule the $\mathbf{K}_i$ operator[2]:

$$\begin{aligned}&\mathbf{K}_i\big(intend_G(rob\_bank) \rightarrow intend_G(buy\_material)\big)\\ &\mathbf{K}_i\big(intend_G(rob\_bank) \rightarrow intend_G(dig\_tunnel)\big)\\ &\mathbf{K}_i\big(intend_G(rob\_bank) \rightarrow intend_G(open\_security\_door)\big).\end{aligned}$$

Consider the sub-goal of buying materials (just the action *buy_material*), which requires (see rule below) an agent which is able and willing to perform the action:

[2] The reasoning activity performed by each agent is driven by its internal deductive engine, that implements deduction w.r.t. a suitable axiomatic system. We refer the reader to [5] for a description an axiomatization of *L-DINF*, that as mentioned extends S5, and for details about its soundness and completeness. For simplicity, we recall here just those axioms that would be used agents to make the inferences pertinent to this example.

$$\mathbf{K}_i\big(intend_G(buy\_material) \rightarrow pref\_do_G(i, buy\_material) \wedge can\_do_G(buy\_material) \wedge do_G(buy\_material)\big).$$

More specifically, the facts $pref\_do_G(i, buy\_material)$, $can\_do_G(buy\_material)$, and $do_G(buy\_material)$ have to be derived by the agent as beliefs in its working memory. In particular, such a derivation would rely on rules and facts included in the long-term memory (and, possibly, on previous contents of the working memory) and, as effect of mental actions, would add the needed beliefs to the working memory.

Let us consider the case of action *buy_material*. Assume that agent $i = 3$ is willing to go and buy material. As mentioned in Sect. 2, the derivation $do_i(buy\_material)$ corresponds to the actual execution of the physical action *buy_material*. However, each mental actions performed to derive $do_i(buy\_material)$ might have a cost, that can be payed either by the agent itself or by the group. Assume that the costs of the last three mental actions (that we call α_1–α_3) that lead to derive $do_i(buy\_material)$ are these:

$$C(i, \alpha_1, w) = 18, \;\; C(i, \alpha_2, w) = 15, \;\; C(i, \alpha_3, w) = 3$$

and that $\alpha_j \in E(i, w)$ for $j \leq 3$, for each world w and each agent i. Assume, moreover, that in the current world, say w_1, the agents have the following budgets to perform mental actions:

$$B(1, w_1) = 11, \;\; B(2, w_1) = 21, \;\; B(3, w_1) = 20$$

Assume that agent 3 is willing to go and buy material in world w_1 and that (because of known beliefs or thanks to previous derivations), in the current model $w_1 \models can\_do_3(buy\text{-}material) \wedge pref\_do_G(3, buy\text{-}material)$. However, agent 3 alone cannot perform the mental actions α_1–α_3, because he does not have enough budget. Nevertheless, the inferential action $[G : \alpha_1]$, with $G = \{1, 2, 3\}$, can be performed, because the other agents can devote part of their budgets to share the cost. Indeed, the group G can perform α_1, because $\frac{C(3,\alpha_1,w_1)}{|G|} \leq \min_{h \in G} B(h, w_1)$. Hence, the belief $do_G(buy\text{-}material)$ can be inferred by each agent. After the execution of the action α_1 the budget of each agent is updated as well. The new budgets, given that we are assuming the policy to divide expenses into equal parts, are:

$$B(1, w_2) = 5, \;\; B(2, w_2) = 15, \;\; B(3, w_2) = 14$$

where we name w_2 the situation reached after executing the action. Analogous steps can be performed by G to jointly efford the cost of the mental action α_2. Finally, action α_3 can be performed by agent 3 alone, in order to obtain $do_3(buy\_material)$.

In similar manner two other derivations can be developed so that agent 1 concludes $do_1(dig\_tunnel)$ and agent 2 concludes $do_2(open\_security\_door)$. This allows the group G to finally joinlty achieve the intended goal.

4 Conclusions

In this paper, we discussed aspects of an epistemic logic, *L-DINF*, that we have proposed and developed to provide a framework for a formal description of cooperative (groups of) agents. We have shown how, in our logic, one can represent situations where the group's behaviour can be customized according to the context at hand, for instance by managing the cost of actions and the preferences/capabilities of agents. Using the short example we discussed about usefulness, simplicity, and expressiveness of *L-DINF* for representing group dynamics of a group of criminals intending to jointly reach a common objective. One relevant aim of this research is, in fact, to build a system that would allow a designer to understand if and how a group of agents might reach its goals. The capability of our logic to provide explainability is a crucial factor, as one might define a Decision-Support System with an *L-DINF* theory as its "core", to allow investigators to model and understand the dynamics of criminal organizations.

Concerning a (future) implementation of such a system, building a suitable friendly user interface would be in order. We may notice that the complexity of *L-DINF*, discussed in [1], is the same as that of other similar logics. Thus, on small/medium instances the implementation might be reasonably efficient.

References

1. Costantini, S., Formisano, A., Pitoni, V.: An epistemic logic for modular development of multi-agent systems. In: Alechina, N., Baldoni, M., Logan, B. (eds.) EMAS 2021, Revised Selected papers. Volume 13190 of LNCS, pp. 72–91. Springer, Berlin (2022)
2. Costantini, S., Pitoni, V.: Towards a logic of "inferable" for self-aware transparent logical agents. In: Musto, C., Magazzeni, D., Ruggieri, S., Semeraro, G. (eds.) Proceeding of XAI.it@AIxIA 2020. Volume 2742 of CEUR Workshop Proceeding, pp. 68-79. CEUR-WS.org (2020)
3. Goldman, A.I.: Theory of mind. In: Margolis, E., Samuels, R., Stich, S.P. (eds.) The Oxford Handbook of Philosophy of Cognitive Science. Vol. 1, pp. 402–424. Oxford University Press (2012)
4. Costantini, S., Formisano, A., Pitoni, V.: An epistemic logic for formalizing group dynamics of agents. Interact. Stud. **23**(3), 391-426. John Benjamins Publishing Company (2022)
5. Costantini, S., Formisano, A., Pitoni, V.: An epistemic logic for multi-agent systems with budget and costs. In: Faber, W., Friedrich, G., Gebser, M., Morak, M. (eds.) Proceedings of JELIA-21. Volume 12678 of LNCS, pp. 101–115. Springer, Berlin (2021)
6. Rao, A.S., Georgeff, M.P.: Modeling rational agents within a BDI architecture. In: Proceedings of (KR'91), pp. 473–484. Morgan Kaufmann (1991). ISBN:1-55860-165-1

Effective Contract-Net-Based Intrusion Detection Using Intelligent Agent-Based Systems

Laszlo Barna Iantovics, Olivér Hornyák, Gloria Cerasela Crişan, Bogdan Crainicu, and Elena Nechita

1 Introduction

An Intrusion Detection System (IDS) is specialized in monitoring system and network activities, and perceiving malicious actions [1]. IDSs usually record data and information related to detected unusual events, and produce some reports. These systems should identify possible incidents and report unwanted attempts. Some organizations use IDSs for other reasons, such as detecting weaknesses of internal security policies.

L. B. Iantovics (✉) · B. Crainicu
George Emil Palade University of Medicine, Pharmacy, Science and Technology of Targu Mureş, 540142 Targu Mureş, Romania
e-mail: barna.iantovics@umfst.ro

B. Crainicu
e-mail: bogdan.crainicu@umfst.ro

O. Hornyák
University of Miskolc, Miskolc, Hungary
e-mail: oliver.hornyak@uni-miskolc.hu

G. C. Crişan · E. Nechita
Vasile Alecsandri University of Bacău, Bacău, Romania
e-mail: ceraselacrisan@ub.ro

E. Nechita
e-mail: enechita@ub.ro

M. E. Cornejo et al. (eds.), *Computational Intelligence and Mathematics for Tackling Complex Problems 5*, Studies in Computational Intelligence 1127,
https://doi.org/10.1007/978-3-031-46979-4_18

The most common limitations of many IDSs include [2]:

- Real attacks that are far below the false-alarm rate are often missed;
- Increased false alarm rate: bad packets generated from software bugs escape and corrupt the Domain Name System data;
- Attacks geared by outdated software versions. The signature of outdated databases can open vulnerabilities to new strategies, unknown at the moment of creation of the database.

The development of efficient IDSs is a very important research direction in assuring the security of the organizations. With the exponential growth of the internet, to offer trustful services became more difficult for the recent IDS systems [3]. *Multics*, developed in 1988, is an intrusion detection and alerting system that operated as an expert system [4]; *IDES* [5] is an audit data analysis and mining system [6], *Bro* [7] is a distributed intrusion detection system prototype that operated as expert system [8]. *OSSEC* represents an example of a host-based intrusion detection system [9].

In this paper, we propose an intelligent decentralized agent-based (the agents are knowledge-based agents) security approach for distributed computer network environments. The proposed IDS is able to allocate tasks to the agents, with the purpose of collecting, analyzing and delivering at the same time data needed for recognizing threats.

2 The Proposed IDS System

Previous network security solutions that we have proposed in the past include several novel cryptanalysis algorithms [10]. Cryptanalysis algorithms offer security solutions to the transmitted data, information and knowledge. A more comprehensive security solution requires an IDS or an Intrusion Detection and Prevention System (IDPS).

Intrusion detection is a difficult problem almost all the time. Some results presented in the scientific literature prove that agent-based systems (sometimes intelligent knowledge-based agents) are appropriate as intrusion detection solutions [11].

In a cooperative multiagent system (CMS) formed by social agents, each agent should undertake one or more roles. The role of an agent defines the contribution of the agent to the problem-solving in the frame of the multiagent system. As an example, we mention a multiagent system formed by agents with master and slave roles. An agent with master role may request to some subordinated agents with slave role to solve some problems. An agent with slave role should solve the problems transmitted by its master.

In the following, we propose a decentralized agent-based security approach for deploying an IDS in a distributed software environment. Some preliminaries of the proposal are presented in [11]. *AbIDS* denotes the proposed system formed by knowledge-based agents. $AbIDS = \{Agent_1, Agent_2, \ldots, Agent_h\}$, where $Agent_1, Agent_2, \ldots, Agent_h$ represent the agents. Each agent is endowed with software sensors (they perceive the software environment and the suspicious activities), software effectors (they execute actions within the environment), autonomy (the agent either is able to fulfill a collaboration request, or it can reject it, due to its autonomy), communication capacity (the agent communicates with other agents or with humans), and cooperation capacity (the agent can cooperate with other agents).

The agents are distributed in the network. Each agent may undertake one or more roles having the commitment of fulfilling tasks corresponding to those role/roles. We consider the following roles in the system: "Collect", "Analyze" and "Act". Each role has defined before some specific tasks that can be performed for supporting the intrusion detection. If an agent overtakes a role, then it can fulfill some specific tasks.

Role($Agent_r$) = "Collect" means that the agent $Agent_r$ is able to collect data. The *Collect* agents form the first level of the *IDS* are statically mapped onto every node of the distributed system. They are dynamically configured by the corresponding node where they reside.

Role($Agent_r$) = "Analyze" means that the agent $Agent_r$ is able to analyze data.

Role($Agent_r$) = "Act" means that the agent $Agent_r$ is able to deliver data.

Role($Agent_r$) = {"Collect", "Analyze", "Act"} means that $Agent_r$ fulfills all the roles in the system.

We call *specialization* a method that allows the identification of a specific type of complex attack. In the system, there exist more specializations (based on different algorithms, like: neural networks, vector support machines, immune systems etc.) for identifying different types of attacks, with different complexity.

Figure 1 illustrates a particular *AbIDS* system. The *AbIDS* system is formed by the agents: $Agent_h$, $Agent_i$, $Agent_j$, $Agent_q$, $Agent_s$, $Agent_g$. $Attack_k$ indicates a specific attack launched by a hacker, sensed by the agent $Agent_h$. The dashed arrows illustrate the cooperation (cooperating implies communication) links between the agents. Figure 1 presents a particular case where each agent is able to communicate with each other. This is not the case of implemented large-scale software environments, where is difficult or even impossible to connect each agent with each other agent.

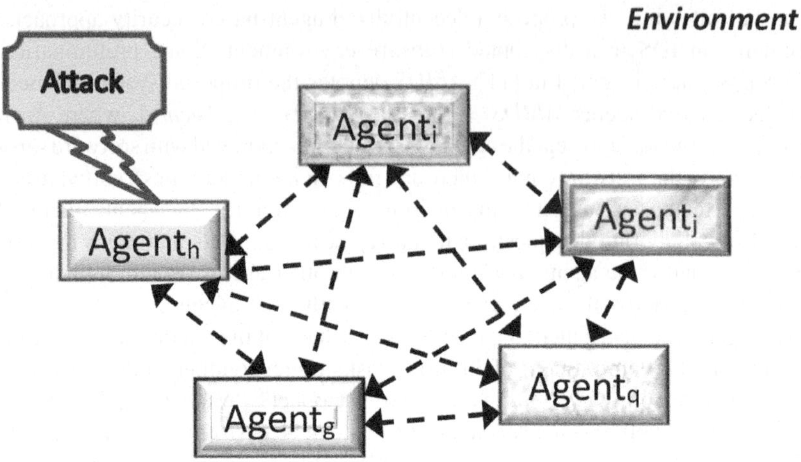

Fig. 1 The proposed IDS placed in an environment with attacks

We call *suspicious* (problem) a situation that seems to be an attack. The algorithm named *Cooperative Problem Solving* describes briefly how the $Agent_h$ proceeds when it receives a PR_k problem. SOL_k—represents the solution of PR_k (either if the alarm was an attack, or not).

A PR_h problem with the AN_h announcement elaborated by the $Agent_h$ agent has the following parameters: $<PID_h| PR_h| SP_h| PAR_h>$. PID_h is the unique identifier (at the system's level) of the PR_h announcement. PID_h (PID_h = <*agent_id*, *problem_id*>) is composed of the *agent_id* (unique identifier of $Agent_h$) and *problem_id* (unique identifier of the problem at the level of the agent $Agent_h$). SP_h is the initially estimated specialization necessary for PR_h solving. PAR_h is a list of parameters that could include different transmitted information, like: eligibility specification (criteria when a problem-solving offer is accepted; the distributed knowledge-based agents may be endowed by different problem-solving methods, and different algorithms such as: immune algorithms, neural networks, support vector machines); bid specification (how an agent should specify its problem-solving capability, for example, estimated accuracy, problem-solving time), and expiration time (the deadline admitted for the problem-solving).

A response $Resp_i$ of an agent $Agent_i$ to a problem announcement AN_h has the following parameters: $<AID_i, OF_i, CB_i, CC_i, SC_i, RV_i>$. AID_i is the unique identifier (at the system's level) of $Resp_i$. AID_i (AID_i = <*agent_id*, "*response*", PID_h>) is composed of the *agent_id* (unique identifier of the agent $Agent_i$ who emitted the offer to the announcement AN_h), the constant "*response*" which specifies that is a response to an announcement, and PID_h (the identifier of the announcement AN_h). OF_i is the bid to the problem solving ("agreement" or "denial"). CB_i is the capability of the agent $Agent_i$. CC_i represents the processing capability of the agent $Agent_i$. SC_i is the estimated specializations by the agent $Agent_i$, necessary for the problem solving specified in the announcement AN_h. The RV_i list values describe how accurate is the estimated necessary specializations list.

Cooperative Problem-Solving Algorithm

IN: PR_k; **OUT**: SOL_k

@$Agent_h$ undertakes PR_k. @$Agent_h$ estimates the necessary processing of PR_k.

If ($Agent_h$ can process PR_k) **then** @$Agent_h$ processes PR_k obtaining the result PR_h.

If (PR_h is the problem solution) **then**

@SOL_k=PR_h.

@SOL_k is verified and transmitted to the PR_k sender.

Else @$Agent_h$ transmits PR_h announcement AN_h to some agents.

EndIf

Else @$Agent_h$ transmits PR_k announcement AN_k to some agents.

EndIf

While (the coming up time to the problem announcement is not ended) **do**

@$Agent_h$ receives and evaluates bids to the announcement.

EndWhile

@$Agent_h$ awards the problem PR_k to a suitable agent $Agent_i$.

@$Agent_h$ receives the result PS from the agent $Agent_i$.

@If is necessary $Agent_h$ can process the result PS obtaining the solution SOL_k.

@SOL_k is verified and transmitted to the PR_k sender.

EndProblemSolving.

There are different kinds of communication and cooperation links between the agents. Equations (1)–(4) illustrate the communication and cooperation links between agents with different roles. There are two types of links: unidirectional links: →, ← and bidirectional links: ↔.

$$\textit{Collect agent}(\text{communicate}) \rightarrow \textit{Analyze agent} \tag{1}$$

$$\textit{Collect agent}(\text{communicate}) \leftrightarrow (\text{communicate})\ \textit{Analyze agent} \tag{2}$$

$$\textit{Analyze agent}(\text{communicate}) \rightarrow \textit{Action agent} \tag{3}$$

$$\textit{Analyze agent} \leftarrow (\text{communicate})\ \textit{Action agent} \tag{4}$$

(1) connectionless, for alerting mechanism—communication initiated by the Collect agent.
(2) connection-oriented, for negotiating the data format collected by the Collect agent and for transmitting data collected by the Collect agent to the Analyze agent–communication initiated by the Analyze agent.
(3) connectionless, for command sending—communication initiated by the Analyze agent.
(4) connectionless, for status reporting—communication initiated by the Action agent.

During configuration, the Collect agents are informed of what kind of data they could collect. A node can host more Collect agents, from which one agent is active at a time. The raw data collected by a Collect agent is arranged in a format previously agreed upon with the Analyze agents, based on a data quality assurance mechanism.

The agents involved in a direct connection/relation by type (1) negotiate the data format. More data formats are possible, but only one communication protocol is to be used between Collect agents and Analyze agents. All data collected by a Collect agent are made available for every Analyze agent, after the first access of a particular data, stored by the Collect agent for a limited period of time. The Collect agents use an alerting mechanism through which the Analyze agents are informed about the availability of newly collected data.

Analyze agents serve as intermediaries between Collect agents and Act agents. The role of the Analyze agents is to decide which information received from Collect agents is important and related to abnormal activities, and then to take the appropriate measures for system protection. All the decisions taken by the Analyze agents are solely based on data received from the Collect agents. Communication between an Analyze and a Collect agent uses a connectionless service for alerting messages (connection initiated by the Collect agent) and a connection-oriented service for *IDS*-specific data collected by the Collect agent (connection initiated only by the Analyze agent). A particular Analyze agent can establish direct connection-oriented relations with more than one Collect agent. The same data could be received by more Analyze agents, which is a very important aspect from the redundancy point of view. At the same time, critical issues such as duplicated data and antagonist decisions are solved together by the Analyze agents in order to have a consistent *IDS* implementation.

Using a connectionless communication, an Analyze agent transmits to the Act agents the precise commands (activating firewalls/filters, dropping packets, closing ports, closing processes, closing active sessions, eliminating users) that the Act agents have to perform. Every action represents an incident response which is a component of a layered security approach mostly established by the Analyze agents.

3 Discussions

It is important to emphasize that Collect agents must commit to gathering a huge volume of data from different network sources. Therefore, these agents will filter, aggregate and pre-process the data before sending it to the Analyze agents. From this point of view, a Collect agent also acts as a classifier of network traffic based on communication protocols and network elements' characteristics.

Receiving data from the Collect agents, the Analyze agents determine if there are suspicious activities. First of all, the Analyze agents perform a classification of security threats and then an analysis of the incident accordingly. Taking into account that Analyze agents receive from the Collect agents already processed information about the network traffic, the Analyze agents use specific attributes to detect traffic anomalies, following a combination of statistical anomaly detection methods, where distance-based outlier detection is the best suitable algorithm [12] in this environment, and capabilities to recognize new anomaly patterns.

In a decentralized architecture, the communication between Collect agents and Analyze agents, and between Analyze agents and Act agents, is carried out through standardized programmable interfaces, which expose the methods of interaction with the agents. One of the major benefits of this approach is that these methods can enable the sharing of knowledge and data synchronization among agents of the same type. This information exchange between agents has a massive impact on the Analyze agents' ability to learn the behavior of new anomalies through neural networks and, in fact, to uncover and formalize unknown, completely different intrusions.

When analyzing the field of threat management, it is mandatory that, in addition to the threat detection mechanism, the prevention component should also be taken into account. An Intrusion Prevention System (IPS) is an active protection entity that prevents security threats in real-time through the detection of malicious activities and, unlike an IDS, it blocks the identified threat. In this context, where agents incorporate programmable interfaces, the proposed solution is able to provide in a scalable and efficient way the input data necessary for an IPS. Collect and Analyze agents can serve and operate, at the same time, both within an IDS and within an IPS, thus providing the agents with a high degree of integrability.

4 Conclusions

There is no unified definition of the intelligence of the agent-based systems. From our point of view, a system's intelligence (intelligence in problem-solving) can be measured according to how autonomously, efficiently, and flexibly the system can solve difficult problems. Complex attacks represent examples of very difficult problems.

In many distributed systems, different types of security solutions for intrusion detection are needed. In this paper, an intelligent agent-based security approach for implementing an IDS in a distributed network environment is presented. A decentralized, agent-based IDS allocates tasks to agents in order to simultaneously collect, analyze, and deliver data needed for identifying threats and for doing effective actions. The advantages of the proposed approach are the scalability, and the ability to cope with network load and latency, without a single point of failure, thus representing a very appropriate IDS model for large and heterogeneous network environments. The property leading to the intelligence of the proposed system is the autonomous, efficient, and flexible collaboration of the agents to identify attacks.

There are many researches [13–17] focused on offering different solutions to intrusion detection systems which proves that the problem of intrusion detection is one of the hottest topics in research. Some intrusion detection systems [18–21] are considered intelligent based on different considerations of the authors. Even trough the development of intelligent systems is very actual there are very few developed universal metrics for measuring the systems intelligence [22–25]. We consider as a very interesting research direction the measuring the intelligent intrusion detection systems intelligence using intelligence metrics.

Acknowledgements This work was developed in the framework of the COST Action. CA17124—Digital forensics: evidence analysis via intelligent systems and practices (DigForASP).

Additional support for the research was assured by CHIST-ERA programme supported through the ERA-NET Cofund funding scheme under the grant agreements, title: Social Network of Machines (SOON), grant of the Romanian National Authority for Scientific Research and Innovation, CCCDI-UEFISCDI, number 101/2019, COFUND-CHIST-ERA-SOON, within PNCDI III. We acknowledge the support offered by the Research Center on Artificial Intelligence, Data Science, and Smart Engineering (ARTEMIS).

References

1. Scarfone, K., Mell, P.: NIST Special Publication 800-94, Guide to Intrusion Detection and Prevention Systems (IDPS), Recommendations of the National Institute of Standards and Technology, Computer Security Division, Information Technology Laboratory, National Institute of Standards and Technology, Gaithersburg, MD, United States, 20899–8930 (2007)
2. Anderson, R.: Security Engineering: A Guide to Building Dependable Distributed Systems. John Wiley & Sons, New York (2001)
3. Kunang, Y.N., Nurmaini, S., Stiawan, D., Suprapto, B.Y.: Attack classification of an intrusion detection system using deep learning and hyperparameter optimization. J. Inf. Secur. Appl. **58**, 102804 (2021)
4. Sebring, M.M., Shellhouse, E., Hanna, M.E., Whitehurst, R.A.: Expert Systems in Intrusion Detection: A Case Study, Proceeding of the 11th National Computer Security Conference, pp. 74–81. Baltimore, MD (1988)
5. Lunt, T.F.: IDES: An intelligent system for detecting intruders,. In: Proceedings of the Symposium on Computer Security; Threats, and Countermeasures; Rome, November 22–23, 1990, 110–121 (1990)
6. Barbara, D., Couto, J., Jajodia, S., Wu, N.: ADAM: detecting intrusions by data mining. In: Proceedings of the IEEE Workshp on Information Assurance and Security, West Point, NY, June 5–6, IEEE CS, 11–16 (2001)

7. Paxson, V.: Bro: a system for detecting network intruders in real-time. Comput. Netw. **31**(23–24), 2435–2463 (1999)
8. Hunter, P.: Distributed intrusion detection systems (DIDS) can make security more adaptive. Netw. Secur. **2003**(3), 16–18 (2003)
9. Ossec documentation http://www.ossec.net/doc. Accessed 23 June 2022
10. Crainicu, B., Iantovics, B.: Cryptanalysis of KSAm-like algorithms. In: Proceedings of the 1st International Conference on Complexity and Intelligence of the Artificial and Natural Complex Systems. Medical Applications of the Complex Systems. Biomedical Computing, B. Iantovics, et al. (Ed.), pp. 130–148. IEEE Computer Society Press (2009)
11. Crainicu, B., Iantovics, B.: An agent-based security approach for intrusion detection systems. In: Hluchý, L., Kurdel, P., Sebestyénová, J. (Eds.) Proceedings of the 7th International Workshop on Grid Computing for Complex Problems, GCCP 2011, October 24–26, 2011 Bratislava, pp. 126–134. (2011)
12. Angiulli, F., Basta, S., Pizzuti, C.: Distance-based detection and prediction of outliers. IEEE Trans. Knowl. Data Eng. **18**(2), 145–160 (2006)
13. Logeswari, G., Bose, S., Anitha, T.: An intrusion detection system for SDN using machine learning. Intell. Autom. Soft Comput. **35**(1), 867–880 (2023)
14. Sivanantham, S., Mohanraj, V., Suresh, Y., Senthilkumar, J.: Association rule mining frequent-pattern-based intrusion detection in network. Comput. Syst. Sci. Eng. **44**(2), 1617–1631 (2023)
15. Karthikeyan, D., Raj, V.M., Senthilkumar, J., Suresh, Y.: Intrusion detection using ensemble wrapper filter based feature selection with stacking model. Intell. Autom. Soft Comput. **35**(1), 645–659 (2023)
16. Amulya, A., Swarup, K.S., Ramanathan, R.: Spectral analysis based robust multi-level intrusion detection in wide area frequency control. Int. J. Electr. Power Energy Syst. **143**, 108430 (2022)
17. Saheed, Y.K., Abiodun, A.I., Misra, S., Holone, M.K.: Colomo-Palacios. R, A machine learning-based intrusion detection for detecting internet of things network attacks **61**(12), 9395–9409 (2022)
18. Gopi, R., Sheeba, R., Anguraj, K., Chelladurai, T., Alshahrani, H.M., Nemri, N., Lamoudan, T., Lamoudan, T.: Intelligent intrusion detection system for industrial internet of things environment. Comput. Syst. Sci. Eng. **44**(2), 1567–1582 (2023)
19. Nagalalli, G., Ravi, G.: A novel MegaBAT optimized intelligent intrusion detection system in wireless sensor networks. Intell. Autom. Soft Comput. **35**(1), 475–490 (2023)
20. Muthanna, M.S.A., Alkanhel, R., Muthanna, A., Rafiq, A., Abdullah, W.A.M.: Towards SDN-enabled, intelligent intrusion detection system for internet of things (IoT). IEEE ACCESS **10**, 22756–22768 (2022)
21. Seth, S., Chahal, K.K., Singh, G.: A novel ensemble framework for an intelligent intrusion detection system. IEEE ACCESS **9**, 138451–138467 (2021)
22. Iantovics, L.B.: Black-box-based mathematical modelling of machine intelligence measuring. Mathematics **9**(6), 681 (2021)
23. Iantovics, L.B., Rotar, C., Niazi, M.A.: MetrIntPair—a novel accurate metric for the comparison of two cooperative multiagent systems intelligence based on paired intelligence measurements. Int. J. Intell. Syst. **33**(3), 463–486 (2018)
24. Iantovics, L.B., Kovacs, L., Rotar, C.: MeasApplInt—a novel intelligence metric for choosing the computing systems able to solve real-life problems with a high intelligence. Appl. Intell. **49**, 3491–3511 (2019)
25. Iantovics, L.B., Emmert-Streib, F., Arik, S.: MetrIntMeas a novel metric for measuring the intelligence of a swarm of cooperating agents. Cogn. Syst. Res. **45**, 17–29 (2017)

Analysing the Influence of the Pandemic on Crime Patterns in North Macedonia

Aleksandra Dedinec, Sonja Filiposka, and Anastas Mishev

1 Introduction

The analysis of crime rates on the national or local level has always been considered as a valuable action that can define the trends in the society and gauge the changes in criminal behaviour as influenced by the social, economic, political events in the region of interest [1]. In the recent years this type of analysis empowered with additional forecasting that can help optimise future use of police resources is increasingly being done using machine learning techniques [2]. This advances in large historical data sets analysis has enabled national police forces to obtain meaningful insight in the recorded events that provide the opportunity for reduced response time by intelligently redistributing the available staff and resources in the areas with highest probability of upcoming crime events.

Taking into consideration that forecasting is based on patterns found in the past events, it is of great interest to analyse how this forecasting is influenced by major events that produce great changes in the general societal behaviour. The recent COVID-19 pandemic is such an occurrence. Recent studies have shown that the pandemic has had a major influence on the newly emerging patterns in crime rates in many nations. However, the most interesting observation is that this influence does not follow the same rules everywhere. For instance, in the US, recent studies [3] show that the pandemic has influenced on considerable drops in the major categories of crimes, while in India the effect has been completely the opposite and the pandemic has triggered a large increase in certain crimes in this country [4]. However, in both cases, deeper analysis show that there is a shift in the types of crimes committed,

Supported by Cost Action CA17124—Dig4ASP.

A. Dedinec (✉) · S. Filiposka · A. Mishev
Faculty of Computer Science and Engineering, Ss. Cyril and Methodius University, Skopje, North Macedonia
e-mail: aleksandra.kanevche@finki.ukim.mk

M. E. Cornejo et al. (eds.), *Computational Intelligence and Mathematics for Tackling Complex Problems 5*, Studies in Computational Intelligence 1127,
https://doi.org/10.1007/978-3-031-46979-4_19

with the pandemic having a large effect on lesser demeanor and little to no effect on serious crime categories. Another question that is being risen is how this shift in criminal behaviour influences the forecasting models in the post-pandemic era? Is the shift going to be a transient behaviour or will it have a long lasting effect on crime rate?

The focus of this paper is to analyse the impact of the pandemic on the crime rates in the case study of North Macedonia aiming to compare the pre- and Covid time periods and find meaningful trends in the changes in criminal behaviour. For these purposes, using machine learning we have curated a data set that contains all police reported crime events in the period of 2011–2021 and applied classification and time series analysis and forecasting to extract and compare the past and current trends in crime rates, emphasizing the most important changes.

The rest of the paper is structured as follows: in section two we discuss the crime events data set generated from public police records, in the third section we provide a general analysis of the data set in the pre- and pandemic period, while in section four we discuss the impact of the pandemic on crime rates modeling and forecasting. Section five concludes the paper.

2 North Macedonia Crime Events Data Set

The Ministry of Internal Affairs of the Republic of North Macedonia publishes an e-bulletin that describes the daily criminal events in the country in natural native language ever since 2011. Using Natural language processing (NLP) on the information that can be extracted from the e-bulletin [5] an initial crime events data set[1] is created where events are classified into broad crime categories, region and geo-location (latitude and longitude). This classification is done using keyword tagging in combination with analysis of the text preceding and following the identified keyword groups. There are 6 general crime categories defined as armed offences, violent crimes, theft and burglary, drug-related crimes, documents/identification-related crimes and traffic accidents. A seventh category—other crimes—is a combination of all reported events that do not fall into these 6 categories including migrants smuggling, fires, cyber bullying, dog attacks and, lately, pandemic-related offences such as breaking curfew, large gatherings, not wearing protective masks and similar offences.

The initial data set has been carefully curated by analysing ambiguous records that do not contain enough information to be automatically classified correctly in order to clean any erroneous data and significantly increase the quality of the data set. In addition to the data, information about the municipality has also been added based on the geographical location in order to obtain a regional view of the information as it is perceived on the level of the local municipality government.

Based on the curated data set, in Fig. 1 the number of crimes per type and year is shown for the span of 11 years that are available in the data set. It is fairly easy

[1] http://crimemap.finki.ukim.mk/home/en.

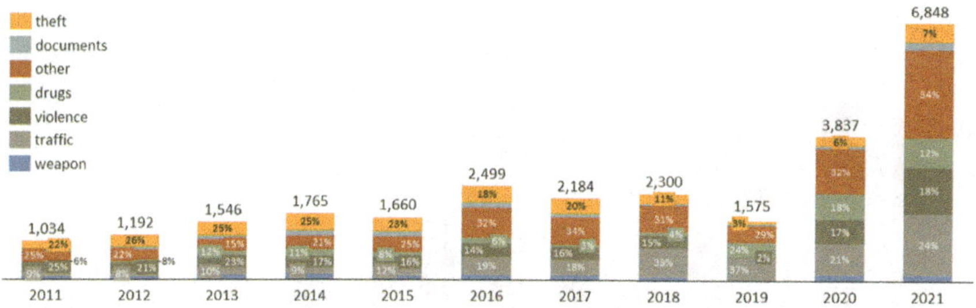

Fig. 1 Number of crimes per type and year

to ascertain that in the years of the COVID-19 pandemic, 2020 and 2021, the crime has seen significant increase in the country almost doubling each year since 2019. In this paper we aim to analyse this information further in order to understand if this is the beginning of a serious trend in rising criminal behaviour and does the pandemic have any direct influence on the recorded events.

3 Crime Data Analysis

The available historical data on crime events in North Macedonia has been used as input to classification and time series analysis. While geographically speaking most of the events are continuously based in the municipalities that make up the general territory of the capital city that daily hosts over half of the country population, from the point of view of types or categories of crime, there seems to be a significant shift in the period before and during the COVID-19 pandemic. In Fig. 2 the change in the distribution of the different crime categories in the before COVID-19 period and during COVID-19 period is shown, as well as the distribution in the overall data-set (including the whole period of 11 years). Please note that in this figure the time period before COVID-19 covers 9 years, from 2011 till 2019, while the during COVID-19 period is a 2 year period from 2020 till 2021. It can be seen that the pandemic has had drastic influence on different crime categories, which is very interesting when compared to the total number of reported crime events presented in Fig. 1 where crime is spiking during the pandemic.

There is major drop in crimes related to documents, such as forgery of documents and false identification, (−46%) which is even a more significant drop when taking into account that part of these events are also pandemic related (i.e. forged certificates of vaccination). Another significant drop in crime is related to armed offences and weapons (−51%). The most significant drop is recorded for theft and burglary (−62%) and these are mostly related to the fact that travel has been very restricted in the period of the pandemic so opportunities for theft of homes have decreased significantly and most of the thefts reported are of stores and individuals on the street. This observed pandemic induced drop seems to be consistent with reports from other

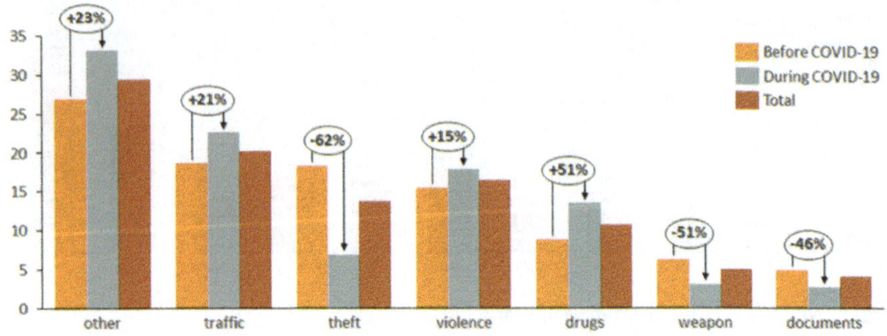

Fig. 2 Change in the distribution of types of crime before and during the COVID-19 crisis

countries such as [7] when it comes to thefts, but the very opposite when considering armed offences where in [7] the authors report significant increase in homicides and gun assaults.

On the other hand, in North Macedonia the pandemic has managed to significantly increase the drugs related crimes (+51%) which is mostly due to the large increase in the number of small drug demeanour (possession, use) and small quantity dealers. Together with the increase in violence related crime (+15%) and traffic accidents and violations (+21%) it can be concluded that the lockdown and social separation has had a tremendous impact on the mental health on people increasing their anxiety and diminishing their tolerance. Again, this observed changes in trends during the pandemic are only partially consistent with other reports, where in drug crime rates are showing significant drops, while violent acts are rising [7].

The crime category that has seen a tremendous rise in the pandemic period is the "other" category that shows a relative increase of 23% compared to the pre-pandemic period. While only 15% of these events are strictly related to breaking the pandemic rules, the same period marks a high rise in number of fire events due to the hot, draughty weather. A particularly interesting case are the increasing events, making 27% of the total other events, that are related with incidents (biting) related to stray dogs. The sharp increase in these types of events is due to the change in regulations related to monetary compensations to the victim payed by the municipality.

Analysing the correlation between the number of active COVID cases and crime events occurrences over time it has been concluded that there is no straightforward relationship between the two. Notable are the short periods of sharp decrease in crime during the strict lockdown in the first half of 2020, which corresponds to reports from other countries such as the UK [8]. However, the resurgence is not gradual in this case, although frequency is increasing over time, especially after the short winding down at the end of the first wave in late 2020.

4 Forecasting Methodology

This paper also analyzes if the pandemic has an impact on prediction whether a certain type of crime will occur on a certain day and in a certain location, based on historical data. For this purpose, crime forecasting is considered as a classification problem in this research and gradient boosted decision trees based models have been developed. Decision trees are one of the basic models in machine learning which partition the data by iteratively asking questions and eventually finding a solution. Gradient boosted decision trees donâŁ™t use just one decision tree, but an ensemble of them. Ensemble learning means combining a number of different models (decision trees in our case), so it tends to be more flexible (reduce bias) and less sensitive to data (reduce variance). There are two popular methods for ensemble learning: bagging and boosting. In this paper, boosting method is used, where the training of the different models is made sequentially (in opposite of the bagging method, in which this is done in parallel). Therefore, gradient boosted decision trees work by building simpler (weak) prediction models sequentially where each model tries to predict the error left over by the previous model. This process is iterative, and, basically in each iteration the model follows the equation:

$$F_{i+1} = F_i - f_i$$

where F_i is the strong model at step i, and f_i is the weak model at step i. At each iteration, a new weak model is trained to predict the "error" of the current strong model. The resulting model aggregates the weak model of each step and thus a strong learner is achieved. The purpose of boosting is to improve the mistakes made by the previous learner (decision tree) through the next learner. In order to predict the error, or the residuals, a loss function is used, such as, mean squared error for a regression problems or logarithmic loss for classification problems. In this paper, the XGBoost implementation of the boosted decision trees in Python have been used, as one of the most popular libraries, due to its wide success in different areas in terms of precision and efficiency.

In the feature calculation process, crime categories are grouped into the following types: violent (which includes violence, weapon and drugs), property (which includes theft) and other (which includes other, documents and traffic). The goal of the developed models is to forecast the violent types of crime, which, as we have previously shown, are affected by the pandemic to a large extent, and where the police should act immediately. Therefore, besides the geographic location which is represented by the longitude and latitude, the following information is calculated and used as input features: number of crimes per type (violent, property, other) in the last 120, 30, 7 days and in the previous day. This means that each sample is characterized with 12 features that include historical crime rates and 2 features for the location.

5 Crime Forecasting Results

Based on the methodology described in the previous section, two models were developed, one for the pre-pandemic period and one for the pandemic period. From these tow models, three case studies are analysed, and a summary of the results is presented in Table 1. In the first two cases, a threshold of 0.38 is set for the fraction of positive labels that we would like to obtain (which is usefully if, for example, the police can patrol only 38% of the analysed regions). When comparing the two cases, although the overall accuracy of the model for the pandemic period is higher, what is more important is that the recall is reduced. That means that, while the police patrols the same percentage of regions, in the corona period 90% of the positive cases will be predicted, while during the pre-pandemic period a higher rate of 94% is correctly predicted. On the other hand, if we set the recall to the same values for the pandemic and pre-pandemic models (case number II and III), we can see that during COVID-19 period the police should patrol an increased number of regions, i.e. 46%, compared to the 38% in the case II. This result is mainly due to the increased rate of violent crimes during the pandemic period. What is also interesting to note, is that 15% of the predicted crimes in the case III will actually happen (the precision) and only 5% of the positive predicted crimes will happen in the case II. The forecasting models are also affected by the pandemic period in terms of feature importance. During the pre-pandemic period, the forecasting of whether crime will happen is mostly based on the corresponding location, while for the pandemic period, the most important features are the sum of other types of crimes in the last 120 days, followed by longitude and the sum of violent types of crimes in the last 120 days. Because of the increased number of crimes, for the pandemic period it is not only the location important, but there is also sufficient historical crime data for the corresponding location that significantly contributes to the predictions.

Table 1 Results of the forecasting models for three case studies

	Case I Pandemic Period			Case II Pre-pandemic Period			Case III Pandemic Period		
	Precision	Recall	f1-score	Precision	Recall	f1-score	Precision	Recall	f1-score
0	0.99	0.66	0.79	1	0.63	0.77	0.99	0.58	0.73
1	0.18	0.9	0.3	0.05	0.94	0.09	0.15	0.94	0.26
Accuracy			0.68			0.64			0.6
Macro avg	0.58	0.78	0.55	0.52	0.78	0.43	0.57	0.76	0.49
Weighted avg	0.93	0.68	0.76	0.98	0.64	0.76	0.93	0.6	0.69
Fraction of positive labels		0.38			0.38			0.46	
Threshold		0.65			0.25			0.52	

6 Conclusion

As a conclusion regarding the ability to predict crime, the results show that with the same percentage of regions covered by the police, in the corona period a lower percentage of positive cases can be detected. In addition, due to the change in crime rates and patterns in the pandemic period, it is important to mention that there is a difference in the developed models for the pre-corona and corona period, in terms of the importance of the input characteristics that affect the accuracy of the predictions.

References

1. Phillips, J., Land, K.C.: The link between unemployment and crime rate fluctuations: an analysis at the county, state, and national levels. Soc. Sci. Res. **41**(3), 681–694 (2012)
2. Chauhan, C., Sehgal, S.: A review: crime analysis using data mining techniques and algorithms. In: 2017 International Conference on Computing, Communication and Automation (ICCCA), pp. 21–25. IEEE (2017)
3. Boman, J.H., Gallupe, O.: Has COVID-19 changed crime? Crime rates in the United States during the pandemic. Am. J. Crim. Justice **45**(4), 537–545 (2020)
4. Bhagtani, H., Roj, S., Raj, K.: Criminal psychology and impact of recent global pandemic: analysis of criminal mind with a sudden surge in crime rate. PalArch's J. Archaeol. Egypt/Egyptol. **17**(6), 10800–10828 (2020)
5. Temelkovski, D., Jovanovik, M., Mishkovski, I., Trajanov, D.: Towards open data in Macedonia: crime map based on ministry of internal affairs' bulletins. In: 9th Conference for Informatics and Information Technology (2012)
6. Kodinariya, T.M., Makwana, P.R.: Review on determining number of cluster in K-Means clustering. Int. J. **1**(6), 90–95 (2013)
7. Rosenfeld, R., Lopez, E.: Pandemic, social unrest, and crime in US Cities. Council on Criminal Justice (2020)
8. Langton, S., Dixon, A., Farrell, G.: Six months in: pandemic crime trends in England and Wales. Crime Sci. **10**(1), 1–16 (2021)

The manufacturer's authorised representative in the EU is Springer Nature Customer Service Centre GmbH, Europaplatz 3, 69115 Heidelberg, Germany. If you have any concerns regarding our products, please contact ProductSafety@springernature.com

Printed and bound by CPI Group (UK) Ltd, Croydon, CR0 4YY
30/09/2026
02225261-0016